Cash for Your Trash

Cash for Your Trash

Scrap Recycling in America

CARL A. ZIMRING

RUTGERS UNIVERSITY PRESS
New Brunswick, New Jersey, and London

Library of Congress Cataloging-in-Publication Data

Zimring, Carl A., 1969–
    Cash for your trash : scrap recycling in America / Carl A. Zimring.
        p.   cm.
    Originally presented as the author's thesis (doctorial)—Carnegie Mellon.
    Includes bibliographical references and index.
    ISBN-13: 978–0–8135–3686–6 (hardcover : alk. paper)
    1. Recycling industry—United States—History. I. Title.
    HD9975.U52Z56   2005
    363.72'82'0973—dc22

                                                                    2005002576

British Cataloguing-in-Publication information is available from the British
Library.

Manufactured in the United States of America

For Maury

Contents

Acknowledgments

I am fortunate to have had the support of several people in the seven years I have worked on this project. At Rutgers University Press, Audra Wolfe was an able and helpful editor who provided excellent suggestions. Also at the Press, Marilyn Campbell, Adi Hovav, and two anonymous reviewers provided logistical help and advice that clarified my thinking. Monica Phillips carefully copyedited the manuscript.

The book is based on a dissertation I wrote at Carnegie Mellon, which provided an outstanding setting to research industrial and environmental history. Words fail to convey how valuable Joel Tarr's guidance on this project has been to me; I cannot imagine pursuing this topic without his advice, kindness, rigor, and support. R. J. Freuhan and Scott Sandage joined Joel in making a diverse committee of readers whose comments enabled me to best articulate my ideas. I thank Scott as well for suggesting the Fats Waller song for the title of this book.

I have been fortunate to receive critiques, probing questions, advice, and assistance from faculty and students at Carnegie Mellon, where Glen Asner, Jared Day, Gerard Fitzgerald, David Hounshell, John Jensen, Eugene Levy, Jim Longhurst, Jason Martinek, Judith Modell, Steve Schlossman, Peter Stearns, and Joe Trotter all made suggestions that have found their way into these pages. Special thanks to Trent Alexander, whose statistical mastery of IPUMS data proved indispensable when writing my second chapter.

Much of this manuscript was revised during a year teaching at Michigan Technological University. I thank Bruce Seely, no stranger to the history of iron and steel, for providing me the opportunity and resources to have a most productive year as I tromped through the snow. My colleagues and students at Oberlin College kept my mind sharp in the advanced stages of revision.

Many industry figures shared their time and memories, and I thank William Breman, Greg Crawford at the Steel Recycling Institute, and the people of the Institute of Scrap Recycling Industries (ISRI). ISRI has a great interest in its members' history, and it provided invaluable information, access to publications, and contacts. I am particularly grateful to Dr. Herschel Cutler and Si Wakesberg, both living archives of information on the industry's past and present, for sharing their time and resources with me.

This research would not have been possible without the financial assistance of several institutions. First and foremost, the U.S. Environmental Protection Agency's Science to Achieve Results (STAR) fellowship program provided two years of support. Thanks especially to STAR program officer Jason Edwards for making sure everything went smoothly for me. In addition, the Hagley Museum and Library, the Baird Society resident scholar program at the Smithsonian Institution Libraries, Carnegie Mellon's Center for Iron and Steelmaking Research, and Department of History provided generous support.

Portions of several chapters were presented as papers delivered at meetings of the American Society of Environmental Historians, the German Historical Institute, the Great Lakes History Conference, the Social Science History Association, and the Urban Historical Association. Feedback from these presentations, especially from commentators Charles Hyde, Martin Melosi, Philip Scranton, and Christine Rosen, influenced the present form of this research. I am also indebted to feedback, comments, and questions from Steve Corey, Kenneth Durr, Paul Gilmore, Ann Greene, James Lide, John Maher, Tom McCarthy, Bill McGowan, Tony Penna, and Tom Zeller at conferences and online. Chapter 3 borrows ideas from an article I wrote for *Environmental History,* and I thank editor Adam Rome and the journal's referees for their suggestions on that piece.

Gail Dickey, Natalie Taylor, Janet Walsh, and the staff of Carnegie Mellon's history department all provided help in the dissertation stage. Carnegie Mellon's Interlibrary Loan Office, under the able watch of

Gerri Kruglak, provided me with scores of materials. The staff, including research librarian Sue Collins, saved me untold troubles.

At the Hagley, Barb Hall, Roger Horowitz, Carol Lockman, Marge McNinch, Michael Nash, Philip Scranton, and Susan Strasser provided a vigorous intellectual climate and guidance through rich archival holdings. Bill Baxter, Ron Brashear, Bonnie Sousa, and Jeffery Stine made my stay at the Smithsonian Institution a productive one. The Photoduplication Service at the Library of Congress, Tom Lisanti at the New York Public Library, and Terri Raburn at the Nebraska Historical Society provided assistance with photographic acquisitions. Mary E. Herbert of the Maryland State Historical Society assisted in copying archival materials. Jason Baldwin and Jen Potter graciously provided photographs. Brenda Snell at Oberlin College provided much appreciated administrative assistance in the final days of revisions.

Friends from both inside and outside academia shared intellectual feedback, moral support, good meals, the occasional place to sleep, and healthy breaks during the writing of this manuscript. Warmest thanks to Trent and Jennifer Alexander, Steve Burnett, Mike Cuccaro, Jyl and Wes Freed, Robin Dearmon Jenkins, Becky Kluchin, Tom Moran, Stephanie Vargo, Paul Nelson, Diana Quinn and too many dogs, Asif Siddiqi, Lisa Sigel, Jon Silver, Erin Snyder, Jeff Suzik, Ursula Rogers-Ingalls Syrova, Laura Watt, David and Elizabeth Wolcott, and my fellow travelers at WRCT for their friendship and generosity.

I thank my parents, Frank and Susan, my stepmother, Michal, and my brother, Dan, for their support over the years. Jen Potter has seen me through almost the entire length of my researching and writing of this project. I can thank Jen for a whole host of reasons, ranging from a loving marriage to showing a true commitment to material reuse around the house, but those salutations would prove inadequate. I thank Jen for being Jen.

Last but by no means least, I dedicate this work to my grandfather. Maury Zimring provided the initial inspiration for this study when we shared several long stories in the summer of 1992 about his father, Abraham Zimring. Abraham was a Jewish immigrant from Austria who, upon his arrival in the United States in 1904, became a scrap peddler in Waterloo, Iowa. My grandfather and his father, and everyone named above, are responsible in countless ways for this book's existence. I alone am responsible for any errors or omissions.

Cash for Your Trash

Introduction

Paul Revere recycled. Readers may be surprised
that one of the heroes of the War of Independence
participated in an activity we associate with the late twentieth century. Revere did not call what he did "recycling"—that term first was
used regularly by the petroleum industry in the 1920s—but he saved
old metal objects for reuse, just as we save cans and bottles today.[1]

Revere did not recycle because he was interested in saving the
environment; I do not know if he had any particular views on his sur-
roundings, other than that they should not be under the authority of
King George. Revere recycled because using old metals provided valuable economic returns. He made his living as a metalsmith, crafting
primarily goods of silver but also copper, iron, and gold. From scrap
metal he forged horseshoes sound enough for him to ride to Lexington. Reuse meant he did not have to purchase new supplies of metal
when he wanted to fashion new horseshoes, or church steeples, or
whatever his customers asked him to make. After the war, these customers included the new government, who purchased some of Revere's
copper for pennies minted during the early 1790s. Subsequently, Revere supplied metals for naval vessels and the dome of the new state-
house for the commonwealth of Massachusetts. Revere collected old
metals from broken farm implements, cookware, and other sources near
his yard in Boston, using them in the production of new goods. His

activities were not unusual at the time; colonial blacksmiths regularly made use of old metal when fashioning new goods.[2]

Today, we see material reuse in a much different way. We recycle because we are concerned about the environmental consequences of our waste disposal. Public programs to collect and process post-consumer recyclables have burgeoned in the United States over the past three decades, ranging from municipal curbside collection pro-grams to office collection programs to mandates by the federal gov-ernment to collect and recycle its own glass, plastic, metal, and paper. Every week, millions of Americans place old cans, bottles, and papers at their curbs for pickup, some even go to recycling centers to drop off various materials. Programs exist for recycling a variety of goods, from automobile batteries to laser-printer toner cartridges.

These programs are common now, for many Americans think re-cycling is a good, even moral behavior protecting the environment. Reusing items tempers our rampant consumption and reduces the amount of garbage we throw into landfills and incinerators. Ameri-cans recycle because they feel doing so is environmentally responsible. They see recycling postconsumer materials as a way of reducing the burdens of consumption by reducing the amount of waste disposed of in landfills. Recycling programs are considered a part of a sustain-able strategy to manage solid wastes.

When curbside recycling programs are successful, these materi-als join a large stream of postconsumer and postindustrial materials—including obsolete industrial machinery, junked automobiles, construction waste, material from demolished buildings, and other sources—first in scrap yards, then in industrial reuse. The activity of reclaiming materials is not new. Reuse of old materials in colonial mills, factories, shops, and homes was widespread. Although reuse practices have changed since then, the republic has never seen a time without substantial quantities of old materials being reused. Much of this ma-terial was reused in the home prior to the late nineteenth century; old clothing was frequently mended and reused as rags or material for quilts. Since colonial times, the urban poor and charity groups such as the Salvation Army collected materials from dumps and city streets for resale to merchants. The people who discarded these materials per-ceived them as waste, yet they had sufficient utility to someone that they were salvaged.[3]

If somehow Revere could travel from 1775 to 2005, he would be perplexed by all of the people taking bins of paper, metal, glass, and materials he would not recognize (plastics) to the front curbs or alleys of their homes, where they would leave them. Eventually, a truck would come along and pick up the bins. The silversmith would see no evidence that the people in the truck had compensated the people in the houses through cash or trade. The arrangement, though its ultimate ends are quite similar to Revere's refashioning of old metal into new objects, would strike him as bizarre.

How and why material reuse in America changed between Revere's time and the present is the subject of this book. What Revere would find odd is what most Americans consider recycling, our standard practice of reusing materials. Many factors have contributed to the changes between Revere's time and today. The advent of heavy industrialization and its production and commodification of waste have had a tremendous effect on the way Americans have treated materials discarded after industry and consumers have used them once. Old metals Revere purchased by the pound were sold by the ton a century after his death.

Perhaps the most perplexing factors in the changing history of material reuse, however, are the complex and often contradictory attitudes the people of the United States have concerning consumption, waste production, and waste management. Technological progress, political change, and cultural attitudes have allowed Americans to distance themselves from the consequences of consumption, resulting in neglect and scorn where those consequences ultimately manifest themselves. Perhaps no "waste" trade (the word, though once universally used, now has pejorative connotations in the scrap recycling industries) has generated as much conflict and recrimination in American society as that of the people who work to return discarded materials to industrial production.

American material reuse since Revere's time has changed because of economic, cultural, and social forces. Susan Strasser chronicles much of this change in *Waste and Want*, in which she argues that Americans went from having a stewardship of objects to ongaging in a throwaway society after the advent of industrialization and mass consumption. The changes in reuse since Revere are revealed in the rise and evolution of the scrap material industry in the two centuries since Revere recast his horseshoes and bowls. The scrap industry to this day exemplifies the processes and contradictions of American attitudes toward their objects and, in particular, the notions of waste relating to their objects.

Two definitions of waste have caused Americans to behave in contradictory manners toward their objects, and the tensions in these contradictory manners have produced a large, vital, and often maligned trade playing upon the different meanings of waste in the United States.

We have always been concerned with waste as pollution, knowing that excrement in our drinking water produces illness, and other matter out of place may be equally destructive. Changing notions of filth, however, produced new disdain of old objects as dirty and produced more postconsumer waste beginning in the late nineteenth century. With few exceptions—most notably in World War II, when war production dictated more extensive conservation measures—American production of waste materials has increased ever since. More waste materials mean more potentially salvageable materials as scrap dealers comb postconsumer and postindustrial discards to bolster their stocks, yet handling wastes, as anthropologist Mary Douglas observed in 1966, requires breaking taboos about cleanliness and order. Individuals who transgress accepted notions of sanitation are themselves unclean and marginal. Douglas argued that a society's construction of pollution "condemns any object or idea likely to confuse or contradict cherished classifications," thus matter disposed of as rubbish is considered unsanitary.[4]

Utility of this matter depends on the social position of the user. In *Rubbish Theory*, Michael Thompson notes that a society's designation of objects as durable goods, transient aging goods, or rubbish of negative worth is dependent on issues of need, order, and context that have class and status connotations.[5] In the United States, cleanliness became a way not only to rid oneself of disease but also to separate oneself from the laborers, indigent, and foreign in the late nineteenth and early twentieth centuries. A good American by the 1920s kept waste away from the home and the body as much as humanly possible.[6]

If the definition of waste as pollution produced more disposal of materials in the United States, a second definition of waste, waste as inefficiency, produced the impulse for increased salvage of disposed materials. Since the late nineteenth century, American industry has concerned itself with fighting what economist Stuart Chase termed "the tragedy of waste." Failing to maximize the value of labor and capital wasted assets. The impulse to maximize efficiency resulted in attempts to identify and use secondary materials when and where they were more affordable than primary materials. Several industries, including steel, paper, petroleum, weapons manufacturing, railroads, and automobile production devised methods to collect, reuse, and purchase salvaged materials, efforts that intensified in the early twentieth century as sophisticated managerial structures emphasized efficiency and sought to avoid lost revenue through waste. This impulse commodified scrap, belying its identity as worthless filth.[7]

The creation of demand for wastes led to opportunities to profit from rummaging through discarded materials, and it is in these opportunities that one can see the cultural constructions that both limited and expanded entry into American entrepreneurialism. These opportunities were open to individuals outside of the dominant culture in the United States because taboos relating to filth precluded most Americans from sorting wastes.

Individuals lacking these taboos who were inclined to start their own businesses found the scrap trade had what sociologist Roger Waldinger calls an opportunity structure that allowed for widespread participation by immigrants. With very little initial investment, a scrap peddler could rise to prominence in scrap within a few years in the fluid period between 1880 and 1920, with many dealers enjoying substantial wealth. Thousands of immigrant-founded scrap businesses established themselves between the Civil War and the end of World War I.[8]

These firms—ones that Philip Scranton identifies as possessing "variety, prowess, and novelty as the keys to profitability and a blending of firm autonomy and judicious associationalism as strategies for growth and for managing competition"—served the needs of large industrial manufacturing by closing the loop of material use. Although large corporations began consolidating the trade in the 1950s, many small, flexible firms remain in business today, and no history of this business can be complete without an understanding of the ways in which small enterprises shaped American patterns of material reuse.[9]

Going into the business involved risks. One was the dangerous aspect of the work; sorting wastes exposed the scrap dealer, and the dealer's workers, to unsanitary conditions and unsorted, potentially hazardous materials. Another risk was stigma associated with the taboo of handling waste. Attitudes toward scrap dealers since the nineteenth century have been frequently negative, xenophobic, and cause for the industry to respond with aggressive public relations efforts that have never eradicated the stigma. Furthermore, the stigma associated with waste, while it allowed many newcomers to take advantage of opportunities shunned by native-born Americans, meant that the effects of waste processing and handling fell disproportionately on individuals and groups who lacked political and economic power.[10]

The definition of waste as inefficiency does not mean that all discarded materials are desired for salvage. The scrap industry has never claimed to eliminate postconsumer and postindustrial wastes, only to reclaim valuable materials, for only those materials coveted by manufacturers return to the cycle of production. Economic and technological changes alter demand for particular materials, creating dynamic markets over time. The adoption of the open-hearth process of steel making at the turn of the century allowed firms to use larger amounts of ferrous scrap without concern for impurities, producing increased demand for iron and steel. Half a century later, adoption of the basic-oxygen steelmaking process (BOSP) reduced demand for scrap sold on the open market as steel producers used more of their self-generated scrap for production. Without demand, recycling does not work and scrap materials lay dormant in yards. Booms and busts for particular materials may be short or lengthy, but demand for these commodities is never static.[11]

The changing demand for materials belies the steady collection of modern recycling programs designed to divert recyclables from the solid waste stream. These programs have an environmental purpose, formed to reduce the amount of waste stored in sanitary landfills or burned in incinerators. Cultural changes in the late twentieth century have made consumer recycling a virtue. Paradoxically, the virtue associated with recycling remains incongruent with the stigma associated with scrap dealers, and public equating of the two is rare. The conflicting images of recyclers for the environment and recyclers for profit may contribute to a lack of understanding—and use—of scrap dealers by recycling programs to maximize the amount of materials returned to industrial production. Recent difficulties encountered by municipal recycling programs indicate that a new understanding of the role of scrap dealers in American material reuse will benefit those dedicated to reducing waste.

Fear of waste, and of the people who handle waste, is a major element of this history. Scrap dealers, the "original recyclers," as they began calling themselves in the 1970s, have had a long and often contentious history in the United States. Scrap dealers have been called filthy, criminal, and even un-American. When I first began this project, several educated people warned me about scrap dealers because (to quote one unnamed acquaintance), "they're all in the mob," a patently false allegation. A long history of ugly caricatures plague the trades, serving to marginalize the work scrap dealers have done for over a century.

This marginalization is unfortunate. When one considers how their work allows others to extend the lives of postconsumer and postindustrial discards, it is downright illogical. Today we celebrate the virtues of recycling. The activity we understand as "recycling" means leaving a cleaner, better planet to our children, and we perceive sorting one's recyclables from garbage as a responsible consumer activity.[12] However, as detractors argue, defining recycling as simply the act of sorting one's discards to save the environment is a fantasy. Criticism of the limits of recycling (made most famous by "Recycling Is Garbage," a 1996 *New York Times Magazine* article by John Tierney) points out that many of the processes involved in converting these collected discards have environmental consequences in the expenditure of energy and the release of pollution into the air and ground. In addition, the programs set up to save the environment fail to grasp the economics of effective material reclamation, ignoring the costs of sorting, transporting, and processing the collected materials, costs often borne by taxpayers in cities with public collection efforts. As Tierney put it, "recycling may be the most wasteful activity in modern America: a waste of time and money, a waste of human and natural resources."[13]

Tierney's criticisms of recycling are, as historian Martin Melosi observes, overstated. Yet Tienrey's general point that economic factors, especially the market for collected materials, are important in assessing the effectiveness of recycling programs reveals the industrial roots of a practice generally seen as environmentally friendly consumer behavior.[14] Demand for the secondary materials by producing industries is crucial to keeping materials out of landfills. Garbage archaeologists William Rathje and Cullen Murphy argue that "recycling has not occurred until the loop is closed: that is, until someone buys (or gets paid to take) the sorted materials, manufactures them into something else, and sells that something back to the public."[15]

Modern consumer recycling is one phase of an activity dependant upon demand for the collected material. Economic utility is a crucial dimension of recycling, and consumers are not alone in valuing the practice of recycling. Industrial producers, especially those who adhere to the discipline of industrial ecology, recognize recycling as a valuable goal. Industrial ecologists argue that what is called postconsumer waste and industrial scrap should not be seen as wastes at all but as potentially valuable raw materials that are being wasted. Industrial ecologists David T. Allen and Nasrin Behmanesh argue that one of the research challenges of the discipline is "to identify productive uses for materials that are currently regarded as wastes, and one of the first steps in meeting this challenge will be to understand the nature of industrial and post-consumer wastes."[16]

The problem of waste is one of an open system in which materials pass from raw materials to manufactured items to consumer items to waste, where it leaves the system. In a closed industrial loop, materials that could be disposed as wastes are instead recycled into the manufacturing process, extending the life cycle of materials and reducing burdens on the environment. Attempts to produce closed industrial loops are crucial to sustainable production and consumption. Industrial ecologists recognize closed loops as the basis for recycling, and current work in the field seeks to maximize the efficiency of such loops in the market economy.

Industrial loops, however, have a history as old as industrial production. In many ways, the loops are as old as civilization, as craftspeople have been reusing old metal, rags, bones, fats, and other materials throughout human history, and as Pierre Desrouchers notes, many practices of industrial ecology are far older than the discipline, ranging from Native American reuse of animal tissues to nineteenth-century industries reusing old metal and rubber. Modern industries, including the chemical and petroleum industries, have understood the value of recycling materials for decades.[17] No industry has been more instrumental to the success of material reuse than the scrap material industry, whose thousands of firms represent the conduit between many recycling pickup programs and the manufacturers who complete the cycle of reuse that makes collection a benefit to the environment.

This book seeks to follow the changing patterns of materials reuse from colonial times to the present through an examination of the scrap recycling industries. The scrap peddlers and yards that developed into a loosely organized industry represent important continuities between Revere's time and today. Scrap dealers' activities achieve many of the goals of modern recycling programs in that they divert materials out of the waste stream and landfills and reduce solid waste disposal. Scrap dealers emphasize their work contributes to a better environment (by saving energy, saving virgin materials, and reducing waste disposal) in industry publicity materials and lobbying efforts before state and federal governments. Yet scrap dealers have encountered conflict when doing their work that reveals tensions in American society about consumption, waste, and ethnicity that have developed since the colonial era.

Colonial metalsmiths recognized the value of reuse, collecting small quantities of metal locally for their own purposes. As the United States industrialized in the nineteenth century, a substantial trade emerged that recognized a systematic, closed loop of production and consumption. Scrap material traders built networks collecting, processing, and selling metals, rags, and other discarded materials that formed the building blocks of industrial production in the nineteenth and early twentieth centuries.

Modern scrap dealers resemble the early peddling operations of the industrialized United States; though the technology involved in the trade has evolved and the economic and legal dimensions of the industry have become more complex, the basic skills essential to the successful scrap dealer have remained the same. These are the ability to find, assess, and sell valuable materials and the ability to cultivate relationships with suppliers and customers. Successful peddlers could forge relationships that produced durable businesses; the Pittsburgh partnership of Mullen & Maloney grew from a peddling operation in 1834 to one of four downtown Pittsburgh scrap firms handling twenty-five thousand tons of scrap iron in 1876. Thousands of similar businesses were established in American cities between the Civil War and the Great Depression.[18]

The activities of nineteenth-century waste material traders conform to our understandings of the raw-material-to-new-product-to-raw-material model, yet that model is inadequate to understand its history. It misses an important cultural dimension to the material life cycle. The materials scrap dealers handled passed through the status of waste, in which it was identified as junk, rags, old metal, or other obsolete materials. It is impossible to understand how scrap recycling developed into an industry, who participated in it, and where the business took place without considering the ways in which notions of waste shaped the labor, location, demographics, and status of the waste trades.

"Waste"—as both a wasting of resources and as material befouling homes, cities, and landscapes—is central to understanding American constructions of material reuse. In the late nineteenth and early twentieth centuries, the creation of waste of all kinds increased dramatically due to industrial production and changing notions of sanitation. As manufacturers demanded larger quantities of secondary materials, the growing industries that developed to manage those wastes—industries including garbage hauling, hazardous waste storage, and scrap material trading—were much maligned. Industrialists, politicians, and even progressive reformers, who in many ways displayed the social concerns associated with contemporary recycling, saw scrap handling as work that was not only physically dirty but also morally degrading. That perception reflected revulsion at the kinds of people who did the work as well as the work itself, conflating the unsanitary status of the work with xenophobic stereotypes of the immigrants willing to do the work. Such disdain ensured that the people who entered the waste trades were not those perceived as respectable members of society. This marginalization is intuitive, for waste is by definition negative.

Changing values of acceptable hygiene and health contributed to modern notions of waste management. As Susan Strasser observes in *Waste and Want: A Social History of Trash*, Americans forsook a "stewardship of objects" in the late nineteenth century, opting for the convenience of disposable mass-produced goods over repairing and reusing old cloths and goods. Women's roles managing old materials in the nineteenth century ranged from patching and mending old clothes to the poor women (and children) who collected rags and old iron in the demolition sites, docks, streets, and dumps of America's largest cities in the nineteenth century. With the advent of mass production and mass consumption, Americans began seeing old or worn objects not as valuable but as waste to be eliminated from the home. Both Strasser and Suellen Hoy have shown that maintaining the order of these things had gender dimensions in the twentieth century, with women expected to maintain the order of things in the household. By 1900, this meant removing dirt and germs from the home to combat disease, and removal of dirt meant removal of old, dirty objects.[19]

As fewer objects were deemed suitable to remain in the sanitized American home, American generation of wastes as garbage increased. Garbage disposal was most conspicuous and problematic in highly populated areas, and cities developed increasingly sophisticated public and private systems to manage their wastes. Martin Melosi's and Joel Tarr's research on urban waste management systems reveals that though there has never been a perfect sink for our trash, American cities have developed elaborate systems to remove waste from residential and industrial areas and put them in the ground, water, or (as smoke) air. Urban living and industrial production produced domestic garbage, human wastes, industrial smoke, and other materials that were seen as nuisances or health threats. Lacking ways to eliminate production of the rejected matter, municipal governments attempted to find sinks to put wastes where they would not affect urban residences. Dumps, incinerators, and, after 1940, sanitary landfills appeared in areas where they would cause the least damage or inconvenience to middle- and upper-class residents, often in declining inner-city neighborhoods and on the urban periphery.[20]

The urban context is important, as the waste trades were centered in cities. Wastes of most kinds, intuitively, amassed in greater volume in densely populated areas, and the manufacturers who desired scrap materials tended to locate in urban areas. Reuse and scrap trading certainly transpired in rural areas; my great-grandfather Abraham Zimring worked as a peddler acquiring broken farm machinery and cookery in and near Waterloo, Iowa. The intensive, specialized processing and trade of scrap materials in large volumes is, however, urban in its orientation, with small and large scrap firms operating near the large industrial interests they supplied.[21] The concentration of scrap trading for much of the nation's history is the urban Northeast, with activity spreading west and south as those regions industrialized. This urban dimension to the trade highlights its relationship with the complex patterns of settlement, consumption, and waste in modern society. The scrap material industries, then, are central to American practices of material reuse since industrialization. The industry's history is particularly informative because it provides a lens to view the major social, economic, technological, cultural, and political dynamics affecting American patterns of material reuse.

1 | Rags and Old Iron

Francis Bannerman III was angry. He had received in September 1872 what he perceived to be very poor quality goods from a Glasgow-based scrap trader named Peter Dixon. Bannerman was a Brooklyn-based businessman who bought and sold a wide range of materials ranging from old iron, lead, and copper to rope, waste paper, and war memorabilia. His complaint to Dixon regarded a shipment of grass rope trimmings and waste paper; Bannerman alleged that the shipment featured a small quantity of good material covering inferior materials damaged by tar and water. Bannerman was concerned, for he intended to sell the paper and rope trimmings to other customers and was now not only short of supplies but had already paid Dixon for the agreed shipment. The two men agreed to work out compensation, and Bannerman subsequently offered to buy more goods from Dixon, albeit "good stock no tarred manilla or wet small stuff," at a price to Bannerman's liking.[1]

A quarrel over the relative quality of what most people would consider waste might seem peculiar, but there was nothing unusual about Dixon and Bannerman's trade and dispute. In 1872, hundreds of small dealers on both sides of the Atlantic made their living collecting, buying, and selling old metals, rags, waste paper, and other scrap materials to each other and to factories. How did this trade in waste materials between two men working thousands of miles apart come about? And why did the two men make such a fuss over the relative quality and value of materials many would see as worthless filth?

The practice of finding and reusing old materials is rooted in a practice as old as civilization itself. Classical civilizations in China, India, and the Mediterranean all collected and reused old materials, particularly metals that could be remelted and refashioned. Biblical verses refer to old iron's use, converting plowshares into swords (Joel 3:10) and swords into plowshares (Isaiah 2:4; Micah 4:3). Poor people, including widows and orphans, often scavenged for subsistence. Such small-scale reuse of old materials was common in early modern Europe, where peddlers roamed villages trading new goods for old in the seventeenth and eighteenth centuries, and it was common in the American colonies, where blacksmiths supplemented their stocks with old iron, silver, copper, and gold.[2]

Dixon and Bannerman's relationship, however, represented a new dimension to this practice, for Dixon was based in Scotland and Bannerman in Brooklyn. Men of means had done business in old materials in colonial times; however, such trade was done to supplement or bolster other lines of work. Those individuals who dealt exclusively in old materials were seen as indigent. Bannerman, though he traded exclusively in old materials and not as a supplement to another vocation, was far from poor, having gone into his father's successful business as a young man. Francis Bannerman II had moved from Scotland to New York in 1854 and started trading old iron in Brooklyn during the Civil War; Francis Bannerman III was heir to a thriving business in old materials. One of the reasons Bannerman was so irritated at Dixon was that the shipment was so poor that he "had for to sell it at 3/8 of a cent a pound below a lot that I got from Magee Son and Co. of Liverpool which cost me the same price as [Dixon's]," eliminating any profit for Bannerman.[3]

Dixon, like Bannerman, had a local business supplying both industrial customers and other scrap dealers, and the correspondence between them was indicative of the regular transatlantic trade in waste materials by the late nineteenth century. That trade had a more recent history, one tied to developments in the industrialization of Europe and the United States. Its origins lie in the changing nature of demand for scrap materials produced by the onset of heavy industry in the nineteenth century. Beginning with paper manufacture, then extending to steel making and other manufacturing processes, industrial producers began to demand unprecedented quantities of scrap materials for mass production. This heightened demand led to new business opportunities for scrap collectors, transforming small, local trades into larger businesses with transatlantic trade patterns between 1800 and 1880.

Material Reuse in Colonial America

The transatlantic trade developed as material use and reuse in America changed. Material reuse was commonplace in the United States in 1800; it was common through the entire colonial period, but the ways in which Americans used and perceived old objects differed from how Francis Bannerman made his living. Most material reuse during the colonial period occurred within the home or in small artisan shops. Consumption patterns prior to heavy industrialization compelled families to extend the lives of the items they owned rather than purchase expensive new items, particularly clothing. Susan Strasser calls the care and attention that Americans took a "stewardship of objects" that defined the household economy before the period of mass consumption. Craft skills, including sewing and knitting, were valued in the household, and women taught their daughters to mend old clothing to aid the household economy. Once clothes outlived attempts at repair, many were reused in the household as cleaning rags or as stuffing for furniture.[4]

Artisans and smiths similarly extended the lives of objects around the shop. Blacksmiths frequently remelted old horseshoes, cookery, and other objects to cast new ones. At the end of the eighteenth century, American scrap trading consisted primarily of blacksmiths acquiring small amounts of discarded metals from local sources in order to repair or supplement existing metal goods. In this way, local smiths melted and recast scrap iron, tin, copper, bronze, lead, and alloys such as brass and pewter. Paul Revere was perhaps the most famous example of a colonial old metals dealer, but blacksmiths and silversmiths in the colonies made common and regular use of old metals in 1775 and had done so for two centuries.[5]

English and colonial blacksmiths used old iron for horseshoes and other goods in the early modern period, getting their metals mostly from local sources.[6] The colonies produced iron from ore as early as 1645; by 1700, American iron accounted for almost 2 percent of world production and about 10 percent of British production. Due to restrictions from the mother country, the colonists could not establish new ironworks to produce finished products, limiting most of the colonial blacksmith's work to mending existing goods. Old iron proved a useful and affordable material in small-scale fashioning and repairing of goods, though impurities in the metal made its use in iron production problematic.[7]

Reusing metals was an old practice among craftsmen, especially those who worked with lead and tin. These metals have melting points low enough that they can be melted over a wood fire. Pewter, an alloy of lead and tin was often remelted and recast, and consumer products made from lead, tin, and pewter proliferated across the country. Peddlers going door-to-door often traded new pewter kitchenware for old.[8] Pewterers also fashioned metal goods out of other metals, especially copper and brass. Though brass had been cast in the colonies as early as 1644, copper was scarce in the United States and braziers relied upon supplies of old metal prior to independence.[9]

Blacksmiths and other craftsmen had uses for scrap materials on a small scale and often accepted scrap in trade. They usually got such scrap from local customers in exchange for finished or repaired products. The scale of American industry and the transportation options available to scrap traders in the early nineteenth century meant that material flows from these activities occurred on a local scale. Local exchanges involved people using handcarts, horses and buggies, and sacks to carry loads of rags and small pieces of light metal scrap.[10] Extralocal trade occurred on a small scale. Advertisements in eighteenth-century newspapers indicate small quantities of old metal (as well as rope, canvas, and rags) were imported to Philadelphia from the Caribbean alongside such common imports as molasses, rum, sugar, European luxury goods, and human slaves. For the most part, the colonial trade in old materials was on a local scale and involved small quantities.[11]

Peddling was a common part of American life at the dawn of the nineteenth century. Peddlers canvassed urban streets and countrysides trading new goods for rags, broken pottery, and other discards that could be refashioned into new products. They employed horse-drawn wagons, pushcarts, or their own feet for transportation, storing goods either on a wagon or in sacks. A critical skill for the peddler was the ability to discern the value of offered scrap materials and then effectively barter new goods for them. Peddlers then traded their collections to local blacksmiths or textile manufacturers for money, completing a cycle of reuse that got old materials out of homes and farms and into places where they could be used to fashion new goods.[12]

Rags for Paper

Several changes in production techniques after 1790 raised manufacturers' demand for old materials, and this change in demand transformed what had been small, localized trading circles into industrial activity spanning two continents separated by the Atlantic Ocean. The expansion of the rag trade was an early example of both the industrial expansion of scrap trading and the changes occurring in demand for other scrap materials in the nineteenth century. Production and distribution of textiles led to a greater supply of textiles consumed and a greater supply of textiles discarded.[13]

The life cycle of textiles did not end when consumers' clothes grew worn. Textiles, especially linen, could be used in paper manufacture. Linen and cotton fibers had been used by European papermakers for centuries. In fact, *Scientific American* dated the first known use of linen rags in papermaking to 1178. Linen rags were particularly valued because they were easily rendered into paper and their fine weave produced a smooth writing surface on a durable paper. Harvesting and milling linen and cotton for papermaking involved labor and mechanical costs; the costs of collecting and sorting rags were less, and the quality of the paper produced was equal and often superior to that made from virgin cotton.[14]

Changing processes in paper manufacture in the nineteenth century led to increased productivity. Between 1820 and 1880, the American paper industry switched from hand-powered mills to mechanical mills, increasing speed and productivity of cheap paper without adding to labor costs. Increased productivity meant growing valuation of rags as affordable and desired raw materials.

Paper manufacture involves breaking down vegetable matter into cellulose fiber, forming it into thin, wet sheets, drying them, and giving them surfaces suitable for writing or printing. Using the hand methods of the early nineteenth century, workers performed each of these steps in a different section of the mill. Workers sorted and prepared rags in the rag room. Power-driven machinery washed and broke down the rags into fiber in the engine room. Craftsmen formed the fiber into sheets of paper in the vat room. Workers hung the sheets up to dry in the dry loft. A workman applied a coating to make the sheets less ab-

Figure 1. "In the Rag Trade," drawn by Arthur Boyd Houghton and engraved by W. L. Thomas. *Courtesy Picture Collection, Branch Libraries, New York Public Library, Astor, Lenox and Tilden Foundations.*

sorbent in the size room. Finally, workers smoothed the paper and prepared it for commercial distribution in the finishing room. The process was slow and labor intensive.[15]

Mechanical methods of papermaking reduced the number of workers needed in the process and increased production of paper. The Fourdrinier papermaking machine, patented by Frenchman Nicholas Louis Robert in 1799, became widely used in English paper mills in the early nineteenth century and in American mills after 1825. In the Fourdinier machine, wet rag stock entered the head box, where it was stretched out, shaken, rolled, and pressed into paper, all by mechanisms inside the machine. Workers in the early mechanized mills cut the reeled paper into sheets, which were carried to lofts for drying. The Fourdinier and similar mechanical methods of fashioning paper reduced the number of workers needed in the papermaking process and increased speed of production. Productivity in the industry increased as a result, as did the demand for rags.[16]

With the proliferation of mechanized papermaking increasing output after 1830, mills intensified their efforts to collect rags. Mills directly employed rag pickers. Papermakers set up rag routes and sent teamsters along them regularly. Village storekeepers often accepted rags in trade for new goods and then sold rags to teamsters who brought the rags back to the mill. A local paper mill sent scavengers to local dumps and peddlers to houses to secure rags, preferably white ones but also discolored rags. The mills paid the peddlers and scavengers for their collections, then processed them in "rag-rooms" usually staffed by female workers.[17]

The flow of rags between 1800 and 1830 was extralocal only to the extent that those new textiles fashioned from rags were exported from the locality; demand was insufficient to warrant the transportation costs necessary for long-distance shipping of rags back into industrial production. Transportation expenses were high because overland transportation in the United States in the early nineteenth century was difficult. As late as 1816, shippers could bring a ton of goods three thousand miles by sea for about nine dollars, but the same sum would move a ton of goods inland by wagon only thirty miles. The construction of turnpikes and canals allowed trade between the North Atlantic mercantile cities and developing cities in the interior between 1815 and 1840. The development of the steamboat and construction of canal systems made the waterways of the interior United States navigable to conduct in the 1820s and 1830s. The United States had more than three thousand miles of canals by 1840. New manufacturing centers in the west developed, including Pittsburgh, Buffalo, and Cincinnati. These cities had access to waterways linking them to a network to the developed cities on the East Coast.[18]

Improvements in transportation allowed easier movement of materials within the United States. Some dealers used canals and rivers to send their goods outside of their local areas and beyond state lines, but most scrap and rag dealing before the Civil War did not cross state lines. Transportation of materials had an indirect effect in the trade. Manufactured goods such as clothing and cooking utensils proliferated in the West and South, giving traders there more available discarded consumer goods to collect and trade.[19] By 1850, mills had expanded their purchases to include a transatlantic trade in which independent brokers collected bulk volumes of rags from southern and eastern Europe and sold them to Berkshire firms.[20]

Once paper mills had collected or purchased their rags, they needed to sort and prepare the cloth for manufacture. Mills sorted rags by fiber, color, and cleanliness, often employing immigrant women to do the sorting. Separation allowed for less beating time, less damage to fibers, and greater savings to mill owners. Rags used in paper production were either cotton or linen, traditional European paper fibers because of their ready availability.[21]

Mechanization in the 1850s led to periodic shortages of rags as increased manufacturing capacity spurred demand beyond the locally available supply. Some mill owners responded by purchasing rags from beyond the local vicinity. Access to the Atlantic Ocean allowed for transportation of rags from Europe without requiring costly inland hauling charges. A trade involving rag dealers from Italy and southern Europe emerged in the 1850s as linen rags in the United States became scarce.[22]

About ninety-eight million pounds of rags were imported in 1850; twenty-five years later the total had increased to 123 million pounds. Most of these rags originated from the United Kingdom and British possessions, with the rest coming from Italy, Austria, Turkey, and Germany, and about twenty other countries.[23] The use of imported rags by American paper mills varied. Some mills relied heavily on imported rags while others used mostly domestic sources. In 1832, 75 percent of the rags used in the mills of Lee, Massachusetts, came from Europe, but papermakers in Dalton used 78 percent domestic rags that year.[24] International trade funneled largely through New York City to the United Kingdom.[25]

As rag importing increased, wholesalers rapidly organized the burgeoning rag trade. Mills and dealers established classifications for rags by the early 1840s. The most common classification was a divide between No. 1 rags and No. 2 rags, with No. 1 rags generally being white and of a finer weave and No. 2 rags being rougher, with some discoloration or other impurities. Color was an important variable; in 1865, rag dealer Robert Howarth placed an advertisement in the *Delaware County American* offering ten cents per pound for white rags and old papers and four cents per pound for colored rags and papers. Grades were not uniform: Identification of No. 1 and No. 2 rags depended on each buyer and seller's interpretations of grades, which could differ pointedly and cause many disputes. Despite significant subjective differences, grading rags expedited the mill owner's task of ordering appropriate rags and controlling rag supplies.[26]

Creating Commodities

As rag collection and purchase became a common part of the paper industry in the middle of the nineteenth century, trade in waste materials grew, partly because more materials were wasted.

The growth of American industrial production involved exploiting vast and abundant supplies of raw materials, supplies so apparently abundant that they seemed inexhaustible in the mid-nineteenth century. Wealth generated by abundance produced increasing amounts of waste in the nineteenth and twentieth centuries, amounts conspicuous to foreign observers. On a visit to San Francisco in the mid-nineteenth century, Japanese educator Yukichi Fukuzawa observed "an enormous waste of iron everywhere. In garbage piles, on the seashores—everywhere—I found lying old oil tins, empty cans, and broken tools. This was remarkable to us, for in Yedo, after a fire, there would appear a swarm of people looking for nails in the ashes."[27]

American manufacturers and consumers behaved as if there was no end to this abundance as the United States expanded after the Civil War. By the end of the century, however, industrialists recognized limits to this abundance. Supplies of affordable raw materials such as cotton, iron, and rubber could not keep up with production. Engineers and mill owners grew concerned over raw material scarcity and resource conservation became a goal of many manufacturing plants. Industry managers viewed waste less as refuse taking up space and more as potentially reusable material. In many cases, what had heretofore been seen as refuse became perceived as a valuable commodity.[28]

Rags represented the most widely coveted materials in this market, but businesses devoted to finding, sorting, buying, and selling rags, old iron, copper, rope, and other materials became conspicuous in American cities. Material reuse was a common feature of both urban and rural life; peddlers traversed both city streets and rural farms for old materials, a practice that would continue well into the twentieth century. Businesses engaged in salvaging old material, however, became predominately urban entities doing business near industrial customers. Though industry could and did situate in rural areas, most mills—and most junk, rag, and old metal businesses—were located in or near the largest cities. The increased demand for rags between 1840 and 1860 combined with continued local demand for old metals to produce a proliferation of businesses devoted to sorting and selling old materials in the United States' largest mercantile cities. Boston featured many such dealers; Philadelphia and especially New York City hosted imports of rags, rope, waste paper, and scrap metals. Business listings in New York City and Philadelphia's city directories indicate a proliferation of firms engaged in trading rags and other old materials in the two decades prior to the Civil War.

By 1860, most of the firms listed in each city identified themselves as junk dealers, with no differentiation between scrap materials. A junk shop might feature an assortment of discarded materials ranging from rags to old metal to discarded consumer products such as furniture; "junk" was an all-encompassing and inclusive term. A minority of sellers identified themselves as specialists in rag dealing. More rare was the dealer who identified himself as handling tin scrap, iron scrap, or other specific materials other than rags. "Scrap" was a word used for describing a specific material that a dealer sold; far less frequently was a dealer in such materials referred to as a "scrap dealer" in the nineteenth century. Materials identified as scrap were just as frequently called "old," such as "old iron" or "old rubber." Trade publications such as *Iron Age* usually identified materials as "old cast iron" or "old wrought iron." By the last quarter of the nineteenth century, the collected trade in old materials was often referred to as the "waste trades." Dealers did not consider "waste" and "junk" pejorative descriptors, at least not enough to prevent their frequent use in advertisements.

The number of junk, rag, and scrap establishments in New York City's directories grew from forty in 1840 to seventy-one in 1851 and to 233 in 1860. The New York firms initially advertised exclusively as junk dealers in 1840. The 1851 and 1860 directories indicated a mix of rag and junk firms operating in the city.[29] Waste-trading firms barely registered in Philadelphia's city directories in 1840, with only four rag businesses listed. Thirty-one rag traders and one junk trader were listed in the 1851 directory; by 1860, a customer could choose among 159 firms. By 1860, Philadelphia's waste traders had shifted from dealing mostly in rags to a mix of rag (sixty-three), scrap (seventeen), and junk (seventy-nine) establishments.[30]

Many of these businesses engaged exclusively in a local trade, but some branched out regionally and even internationally. New England dealer Morillo Noyes bought a variety of old materials in the Northeast, including old rubber for seventeen to twenty-five cents per pound in 1858. Noyes employed peddlers outside of his local region to conduct business; in 1854 he had a peddler in New York purchase copper at twenty-two cents a pound and brass at fifteen cents a pound.[31]

Scrap trading gradually moved beyond local borders to include a transatlantic dimension as well as regional movement along waterways. Transportation of materials north and south was hindered by the outbreak of the Civil War, yet the war and the industrial developments that followed brought great benefits to scrap dealers. The Civil War, as with wars past and future, encouraged reuse of finite resources in military applications, lending greater importance to the identification and collection of usable old materials. The Confederate states were not as industrialized as their Union counterparts and therefore resorted to more extensive salvaging to supply their mills and military. Mills found supplies of leather, metals, dyes, and oils were depleted by 1862. Machines from mills closed or destroyed in the war were prized to replace worn machinery in nearby mills. After the Belleville factory in Georgia was burned down for the second time in 1862, its owner George Schley observed several bidders from other mills, and the Confederate military came to purchase the salvageable remains, which included "a large lot of scrap iron, boiler iron, steam boilers, shaftings, couplings & hangers, leather belting, gas pipes, spindle steel, and three large copper kettles." Southern industrialists sometimes used salvaged parts to build new factories. William Amis built a cotton mill from salvaged parts in Carroll County, Georgia, and put it to work weaving yarns for the Confederate Commissary Department.[32]

One could interpret southern salvage operations as a symptom of a faltering war effort, but both the Union and Confederacy increased conservation and reuse of metals during the war. Prices of old lead and iron rose during the early 1860s in New York City, Boston, and Philadelphia. Lead was melted and reused as ammunition. Rags were refashioned into clothing. Gun works used scrap iron to cast new guns. Old became new, and discards became important to the war effort on both sides.

Industry continued to reuse materials at the end of the Civil War. The postbellum period ushered in a new era of industrial development in the United States. Spurred by investments in transportation that had begun three decades earlier, as well as federal legislation to encourage industrial production, such as protective tariffs and encouragement of technological innovations through the National Academy of Sciences, the second half of the nineteenth century saw unprecedented expansion of the nation's productive capacity. The period also saw unprecedented use of raw materials by the nation's growing industries.

Technological changes in productive industries allowed for more extensive use of scrap materials. The paper industry's mechanized processes spurred demand for rags in the early nineteenth century, and similar changes in the steel industry produced new demand for old metals in the final quarter of the century. By 1900, old iron was second to rags as the most widely traded waste commodity in the United States.

The rag trade flourished after the war. In 1866, the price of cotton and linen rags more than quadrupled due to scarcity caused by the war. The domestic supply could not keep up with demand from the paper mills, leading to increased imports from Europe, especially Italy. *Scientific American* opined that "valuable as rags are to the professional beggar, and important as they may be to abject poverty, they are far more important to the world at large; for up to the present time no other material has been found to usurp their place as the basis for paper."[33]

The paper industry continued to use rags as raw materials throughout the century. At the end of the century, technological changes allowed the use of wood pulp, which was high in cellulose fiber and much cheaper and more abundant than cotton and linen rags, in paper manufacture. The use of wood pulp created demand for another material, and subsequently rag traders also dealt in wood pulp. The market for linen rags began to decline for the same reason it had grown; technological change provided for the use of a more affordable material. Dealers adapted by expanding their trade in wood pulp, and many dealers traded in a combination of rags, wood pulp, and old iron even after World War I.

Concurrent to the growing demands of the paper industry, growth in the iron and steel industries in both the United States and Britain produced demands for vast quantities of coal and virgin iron ore. Mills seeking raw materials that were more affordable than virgin ore, the costs of which included mining it from the earth, sought scrap iron as an alternative raw material. Methods of production used in the mid-nineteenth century did not allow for extensive use of scrap. The Bessemer process, the process used most widely in the United States in the late nineteenth century, had use for a limited amount of scrap iron, up to about 10 percent of the charge in order to regulate temperature.

A second method of steel making that allowed the use of greater amounts of scrap iron began to make inroads in the late nineteenth century. Open-hearth furnaces operated at higher temperatures than Bessemer converters did, and they burned off phosphorus and impurities found in scrap that Bessemer converters could not remove. Open-hearth furnaces thus made purified steel from a greater range of materials than was previously possible. The open-hearth process accepted up to 90 percent scrap in the total charge, though a ratio of 50 percent scrap and 50 percent molten pig iron was more common.

Mills could use more scrap in the open-hearth process due to the greater amount of control the operator had over materials used in the heat. The Bessemer process produced steel in a very short time, about twelve minutes, by a violent reaction caused by the oxygen in the air combining with the carbon from the molten pig iron. The combination produced tremendous flames, making it impossible to take a sample to determine the chemical composition of the steel during the process. The operator determined when the steel was adequately refined by looking at the flame's color. The open-hearth process was a much slower operation than the Bessemer process, taking about six hours to complete. The steel cooked in a large oven-shaped furnace and the operator dipped a ladle into the steel to take periodic samples for analysis, adjusting the temperature or ratio of materials in the process to achieve desired specifications.

Technological change in the steel industry would have crucial implications for the scrap iron trade, but the largest changes would be after the turn of the century. The open-hearth process of steel making made slow progress relative to the Bessemer process. Open-hearth production was less than 10,000 net tons in 1875 and failed to reach 100,000 net tons annually until 1880. The Bessemer process remained the main source of steel production in the United States at the end of the nineteenth century, in part because it was the dominant method used to make rails—although makers of structural steel favored open-hearth production. Its limited application of scrap iron was the primary source of demand for the old metal.[34]

The open-hearth process had implications for reuse of other scrap metals. Coppersmiths found the process produced stronger copper and allowed for the use of greater quantities of scrap copper in fabrication. Though the volume of scrap copper traded was a fraction of the iron trade, the use of scrap metal had important implications for industries reliant upon copper. Even before the advent of open-hearth production allowed for reliable reuse of the metal, prices of old copper were high relative to other scrap materials. Copper was a valuable metal made even more so by a scarcity of ore in the United States. Industrial production transformed copper's uses from producing buttons, pins, and sculpture to creating heavy industrial goods. For example, plumbers used copper for pipes, and shipbuilders sheathed the bottoms of wooden ships with copper beginning in the eighteenth century. Copper's conductivity and light weight led to its wide use in batteries and wiring, with copper wiring becoming a common feature of residences, office buildings, vehicles, and industrial plants as they came to rely on electricity. By the 1870s, copper mining expanded to Chile, Spain, Cuba, and the upper peninsula of Michigan. By the end of the century, American copper mining had expanded to the west and southwest, and mining in Africa and Australia had expanded.[35] Price was exacerbated by demand as world consumption of copper rose from 50,000 tons in 1850 to 500,000 tons per year at the end of the century. Copper mills applied open-hearth production in the late nineteenth century, allowing scrap copper to be used with the same level of supervision possible in open-hearth steel making and extending the supplies of this scarce and valuable metal.[36]

Other materials witnessed a transformation in their demand as industrial mills joined local craftsmen as scrap consumers. As rubber became a more widely used ingredient in manufacturing shoes, clothing, tires, and industrial machinery, the value of scrap rubber increased. Thomas Hancock established England's first rubber factory in 1820. Because the raw material was valuable and had to be imported from countries in warm areas, Hancock reused scraps and trimmings from his shop floor. Other sources of scrap rubber included boot and shoe soles and rubber used in clothing. Charles Goodyear's innovation in 1839 of vulcanizing rubber by heating it with sulfur and white lead made rubber tougher and more suitable for many industrial applications, resulting in an increased demand for natural crude rubber. By the dawn of the Civil War, crude rubber's price trebled. Manufacturers, seeking affordable alternatives, used unvulcanized waste rubber that could be reclaimed.[37]

In the 1850s, manufacturers such as Charles Goodyear and Alva Goodrich De Wolfe ground rubber scraps and pressed them into new rubber goods such as buttons. In the 1860s, manufacturers used acids on rags to separate unvulcanized rubber from cotton and wool where rubber was incorporated into clothing.[38] Manufacturers reclaimed rubber in the nineteenth century by mechanical means. Junk-shop owners and peddlers canvassing urban neighborhoods collected old shoes, boots, and clothes for rubber and sold them to mills. The mills employed women who worked from home stripping rubber from cloth. Manufacturers combined ground-up old rubber with new rubber and other materials to make fillers of various kinds.[39]

Some postconsumer rubber could not be reclaimed because of technological limitations. Mills had difficulty reusing vulcanized rubber, and it was not a source of recycled rubber in the nineteenth century.

The commercially efficient alkali process of devulcanizing rubber was not patented until 1899. Without it, little means for devulcanizing rubber existed, and old vulcanized rubber could not be widely used in industrial production.[40] However, rubber used in boots and clothing could be used to produce vulcanized rubber, allowing unvulcanized rubber two cycles through the industrial process. In 1870, old rubber sold for about fifty cents a pound in the United States. Consumption of old rubber increased during the mid-1860s as a worldwide shortage of fresh rubber increased demand for old material.[41]

Technological changes in the nation's infrastructure allowed expanded trade of old materials, and industrial development in the United States was spurred by improvements in transportation and communication over the second half of the nineteenth century. These improvements benefited trade in rags and old metals just as they benefited heavy industry. Overland transportation, which had been prohibitively expensive to justify much intercity trade of waste materials in the 1840s became more affordable with the building of railroads at midcentury. By 1860, the United States had more than thirty thousand miles of track connecting Chicago, Milwaukee, St. Louis, Memphis, New Orleans, Atlanta, Chattanooga, and other cities to the Northeast. Railroads offered year-round reliability, speed, and capacity to ship goods at high volume. The U.S. postal system and the telegraph allowed coast-to-coast communication by 1861, and over the following four decades, the telegraph and telephone allowed businesses access to rapid, reliable information, increasing the volume of commodities traded across the country.[42]

The rise of investment capital and new systems of transportation and communication allowed giant corporations to grow. Two of the most dramatic examples were Carnegie Steel and Standard Oil, but many industries, from rubber producers to meatpackers, saw large concerns develop during this period. As the scale of production increased over the second half of the century, demand for virgin and secondary materials alike multiplied, causing renewed extraction in mines and forests and more collecting and dealing of rags and old iron.

As transportation and communications systems in the United States developed, options to exchange scrap materials increased. As with general industrial expansion, scrap trading moved from the East Coast's mercantile cities westward. Use of canals and railroads in the Northeast allowed industrial products from Buffalo and western New York to travel east. Scrap materials could travel east and west (as well as north and south via the Mississippi River and railroads); they could also be used locally as demand from growing local industries grew. Buffalo's largest scrap dealer was the Hofeller firm, which took shipments from area farmers for rags, rubber, metals, used bags, and other materials. Hofeller advertised itself as the largest scrap rubber dealer in the world at the turn of the century and delivered materials to Ohio, Pennsylvania, and across the Atlantic to the United Kingdom.[43]

Railroads facilitated westward expansion of industry, allowing the cheapest transport of goods across land. After the Civil War, industrial cities in the Midwest developed, including diversified economies in Chicago and Detroit and single-industry mill towns in western Pennsylvania and Ohio. Notable among the latter were the concentration of rubber concerns in Akron and iron- and steelworks in Pittsburgh, Youngstown, and smaller towns in the region. Chicago became a center for processing, marketing, and distributing goods fashioned from the resources of its vast hinterland. Commodities distributed through Chicago included meat, lumber, coal, and iron, the latter two in sufficient quantities to allow the metropolitan area to sustain steel manufacturing output at the end of the century that rivaled the output of Buffalo, Cleveland, and Pittsburgh.[44]

In addition to supplying transportation that could facilitate scrap dealing over longer distances, railroads also acted as suppliers and customers of scrap dealers. Old tracks and engines were made of heavy iron and became valued commodities for railroads. Trade journals reported on the sale and conservation of railroad equipment.[45] Steel mills collected in-house scrap (also known as "prompt industrial scrap" or "home scrap") in order to eliminate waste and use the cheapest available materials for production.

As demand for scrap materials increased, so did the presence of independent dealers such as Francis Bannerman and Sons. These firms, not beholden to one customer, though very often dealing with one or two large customers, scavenged materials from dumps and demolition sites or purchased them from other scavengers. Scrap firms sometimes employed peddlers, but they often relied solely on the skills of the owner to sort and evaluate scrap from refuse.

In addition to making rounds in residential neighborhoods, dealers purchased scrap from industrial firms that perceived of no use for their wastes. Railroads were large suppliers of scrap iron at midcentury; their discarded cars, engines, and especially old rails provided dealers with heavy, good quality iron to sell to mills and to other railroads.

Ships, including those of the U.S. Navy, similarly provided dealers with large amounts of iron and steel. The Navy frequently auctioned off scrapped ships, a system that removed large loads of scrap from its property. Dealers were responsible for hauling the scrap away, dismantling it for its sellable assets, and disposing of any extant waste.

Scrap collection and sorting was hard work. Scrap metal did not come in neat, symmetrical sizes and shapes; it came in whatever form it was in when it was discarded. These forms ranged from large ocean-going ships that were no longer seaworthy to bundles of wires enmeshed with other debris. In order to sell these materials, dealers had to convert them into portable shapes and sizes of material free from impurities. Ship hulls could be encased in rust that had to be scraped or cut off; potentially valuable copper wires had to be separated from worthless refuse.

The work was not only hard but also dangerous. The two most common ways of processing metal scrap involved cutting it with shears or acetylene torches, either of which could easily maim a person. Sharp or jagged edges of scrap metal could cut flesh and, if rusty, cause tetanus. Rags separated by hand were often stained or contaminated with disease-causing germs, and stored rags provided fuel for fires.[46]

Scrap collecting and processing was the kind of low-status, high-risk work sociologist Stuart E. Perry calls "dirty work."[47] Most everyone associated with the waste trades had to handle "dirty" materials. Nineteenth-century firms were small operations, and shop owners, peddlers, and scavengers all performed physical labor. Historian Philip Scranton characterized small firms of the late nineteenth century as flexible businesses structured in a variety of ways and responsible for much of the innovation in business at the turn of the century.[48] Scrap firms, while affected by the rise of large industrial producers that would become customers for scrap material, did not themselves become large firms with departments supervised by managers. Most had only a handful of laborers (usually under a dozen seasonal employees), and many were one-man operations.

Little investment capital was needed to enter the scrap trade. Since the work was dirty, dangerous, and low status, few natives with other prospects chose to perform it for any length of time. The low starting costs, combined with a lack of competition from established natives, made it possible for immigrants to gain footholds in the trade by becoming peddlers or yard labor, then graduating to store or yard owners.

Francis Bannerman and Sons was one of many small businesses

that flourished in the urban Northeast after the Civil War. Francis Bannerman Jr. developed a small business in Brooklyn in the 1850s. War demand for scrap materials grew the business, and by the time Francis III began dealing and quarrelling with Peter Dixon in the 1870s, the firm had developed a large clientele throughout the New York City metropolitan area and the United Kingdom. Locally, Frances Bannerman and Sons participated in auctions of military equipment conducted by the U.S. Navy selling scrapped ships and weaponry to collectors and mills.[49]

The family business prospered. By 1890, the firm's focus shifted from selling scrap to manufacturers to selling war artifacts to collectors. The firm became the world's largest buyer of surplus military equipment, and collectors around the world relied upon the Bannerman catalog well into the 1960s. Bannerman's concerns regarding the quality and quantity of the materials he purchased indicated some of the risks a buyer might incur, yet the firm itself did not avoid suspicion. In the early 1900s, Bannerman sold supposedly authenticated links of the chain used by American forces to block British entry to the Hudson River. The links, Bannerman alleged, were moved by the Navy to the Brooklyn Navy Yard, where first John Abbey and then Bannerman began selling links for as much as $350 apiece. The links wound up in museums, private collections, and even cemeteries. Decades later, engineers established that the links were British in origin and in no way resembled the actual chain's links. Inaccurate information remained an issue in the old materials trade, something Bannerman was all too aware of from his dispute with Peter Dixon.[50]

Francis Bannerman and Sons' growth in the late nineteenth century was part of an expansion of the waste trades in the nation's mercantile centers. New York City and Philadelphia's directories indicate the proliferation between 1860 and 1890. In 1866, *Trow's New York City Directory* listed 250 junk dealers, 132 rag dealers, and 1 scrap dealer. In 1880, *Trow's* listed 333 junk dealers, 139 rag dealers, and 1 scrap dealer. The 1890 directory showed a decline in the number of firms—220 junk dealers, 134 rag dealers, and 2 scrap dealers—but the sheer number of firms remained much higher than the 173 junk dealers, 52 rag dealers, and 8 scrap dealers listed in *Gopsill's Philadelphia, Pennsylvania General and Business Directory for 1860*. The 1870 directory listed 117 junk dealers, 78 rag dealers, and 2 scrap dealers, for a total of 197 businesses. Ten years later, *Gopsill's* listed 297 businesses— 114 junk dealers, 181 rag dealers, and 2 scrap dealers—and the 1890 directory listed a similar number—126 junk dealers, 173 rag dealers, and 1 scrap dealer.[51] Dealers enjoyed similar levels of prosperity in Philadelphia. In 1860, 79 junk dealers, 63 rag dealers, and 17 scrap dealers were listed in the city directory, indicating that the rag trade was particularly vibrant in Philadelphia at a time when the move to wood pulp was beginning to transform material demand in the paper industry.[52]

Industrial development was far less extensive in the postbellum South, leading to a smaller demand for scrap materials. The South lacked investment capital and a well-developed transportation system and relied on an economic and social system based on agricultural production of cotton. The South hosted some industrial activity during this period; southern businessmen built textile mills in the Piedmont, stretching from southern Virginia through the Carolinas into northern Georgia and Alabama in the 1870s and 1880s. Other investors, noting the deposits of coal and iron present in West Virginia, eastern Kentucky, eastern Tennessee, and northern Alabama, opened mines across the region and established irons mills in Chattanooga and most notably in Birmingham, Alabama. The Tennessee Coal and Iron Railroad Company established Birmingham's largest mill and later sold it to U.S. Steel; its production, along with that of smaller mills, made Birmingham a smaller, southern version of Pittsburgh at the end of the century.[53]

Scrap dealers were not as prevalent in the South as they were in the more urban, industrial, and commercial North, but they did serve local industry. European immigrants were scrap dealers in the South as well as the North. One pair of dealers migrated from Germany to Tennessee through the North during the Civil War. The Union army recruited German Jewish immigrants Nathan Cline and Louis Bernheim in New York City in 1862 and sent them south to fight the Confederacy. By the time they reached Nashville, their inability to speak or understand English caused the army to discharge them. Stranded in Nashville and lacking contacts, prospects, and capital, the two became peddlers. Cline and Bernheim bartered with both northern and southern forces, amassing a large store of rags, paper, iron, feathers, and hides by 1863. Scrap metal from the battlefields was both abundant and in demand, and the business prospered. In the decades after the war, Cline and Bernheim's business expanded as they shipped scrap by barge to Saint Louis, Louisville, and Cincinnati. Cline and Bernheim's experience indicates that though southern scrap dealers were not as numerous as their northern counterparts, firms could flourish south of the Mason-Dixon line.[54]

Easing of federal duties contributed to increases in scrap iron imports in the late nineteenth century. Duties on old or scrap iron coming to the United States date back to 1832, when Congress levied a duty of $12.50 per ton. Congress reduced the duty to $10.00 per ton in 1842 and then changed it to a percent of the total value of the scrap, starting at 30 percent in 1846 and then 24 percent in 1857. As the Civil War increased demand for scrap iron, Congress changed the duty in 1861 to $6.00 per ton and then to $8.00 per ton in 1865.[55]

Attention to imports of old iron and steel indicates the growing importance of ferrous scrap. This was obvious during the war and was notable again in the 1880s. The Bessemer process remained the most widely used steel fabrication process until the turn of the century, but the number of open-hearth furnaces grew steadily in the 1880s and 1890s. Imports of scrap iron and steel escalated in the late 1880s as the expansion of open-hearth production in the United States spurred increased use of all iron and steel, especially scrap. Total iron and steel imports increased by two and a half times between 1884 and 1887, from 733,260 net tons in 1884 to 1,997,211 net tons in 1887. In that same period, ferrous scrap imports increased more than tenfold. The American Iron and Steel Association separated scrap iron and scrap steel into two different categories, presumably using the common definitions of the late 1880s to indicate steel as iron that had been melted completely. The amount of scrap iron imported was 30,192 net tons in 1884 and 351,028 net tons in 1887. Scrap steel imports rose from 8,388 net tons in 1884 to 29,716 net tons in 1887.[56]

Scrap markets were volatile due to fluctuating demand, available transportation, and quality of supplies on hand. Material prices fluctuated from week to week, for scrap materials were commodities, with prices subject to variations in supply and demand. Technological developments might increase applications of and demand for a scrap material, such as the open-hearth process did with scrap iron, and other technological developments, such as developments in mining that made virgin ore cheaper to purchase, might negatively affect the market for scrap. Supplies on hand varied with ability to sell quickly and ability to collect from peddlers, auctions, and other dealers.

Timing sales and purchases of scrap materials was crucial to a dealer's success or failure. Prices of No. 1 scrap wrought iron, for example, fluctuated throughout 1873, but a dealer who bought it at sixty dollars per ton in February and did not sell it soon after was stuck with supplies valued at far less. In May, No. 1 wrought iron sold at fifty dollars per ton. By June, *Iron Age* reported almost no scrap iron transactions in New York and an estimated price of forty-five dollars per ton. The depression that affected sales throughout the iron industry depressed scrap prices to thirty-five dollars per ton by November before rebounding to forty dollars per ton by year's end. Within that period were weeks of irregular prices, with trends changing by the day. Some weeks the trade stood at a standstill as dealers refused to sell at depressed prices and demand from mills slowed as demand for scrap commodities declined. Scrap dealers had to negotiate the volatility of the market successfully in order to remain in business.[57]

Uncertainty in the market led to accusations during periods of heavy demand that dealers were hoarding stock in order to drive prices up. Speculation occurred, although dealers who withheld stock risked being undercut by competitors who could steal their customers with lower prices. Speculating dealers risked being stuck with large stocks on hand when the market suffered downturns.

In addition to worrying about price fluctuations, dealers had to contend with the perception that their materials were undesirable wastes. Definitions of waste varied; a material that could be sold had value to someone, but its value might be suspected or questioned by many. Material designated as old or scrap or waste was rendered pollution in the eyes of some, either because they could not see its utility or because they viewed material put into a junk pile as having been rendered unfit to touch.

The kinds of materials considered junk began to increase after the Civil War. Changes in definitions of hygiene and availability of new objects at affordable prices began to reduce the use of old materials in the American home. The most evident changes in the nineteenth century concerned rags. Early in the century, concern about urban epidemics led to a professionalization of public health officials charged with reducing the risk to public health. Using a miasma theory of disease, they developed measures to reduce dirty materials from house and city. Old, discolored, smelly rags were now seen as a public health threat and a nuisance. During epidemics, rags were impounded and quarantined to prevent disease outbreaks. As the germ theory of disease transmission displaced miasma theory at the end of the century, public health measures continued to attempt to restrict or eliminate rags from contact with most residents. Old objects, then, were waste in American homes, of negative worth, fit for expulsion to streets and dumps as solid waste. Yet at the same time old objects' worth in the household declined, their value in industry rose, compelling agents for mills and factories to seek out discarded rags and iron.[58]

The changing steel industry had a complex relationship with old iron. Mills employing the Bessemer process had to be concerned with the amount of scrap iron in their charges; too much could compromise the quality of the finished product. The market for old materials in 1890 was far larger than it had been in 1840, yet some manufacturers and customers were uncomfortable about scrap's use in industrial production. Woodin and Little Pump House, a manufacturer of engines for industrial use, boasted that its products were of high quality due to the careful selection of metals used in its foundry. "A careful analysis is made of all consignments of iron before used and the low grade and unfit promptly rejected, and we use *no scrap iron* [emphasis original]."[59] Metallurgists shared—to a degree—Woodin and Little Pump House's disdain of scrap iron. *Scientific American* asserted that "no matter how many times it is remelted and processed, inferior grades of scrap iron would produce inferior forgings. Care in sorting good scrap from bad is critical to achieving good forgings."[60]

A scrap iron user needed to display skill in selecting grades of material suitable for production. Scrap metal dealers and buyers adopted the rag trade's classification of No. 1 and No. 2 grades of material by 1870. No. 1 grade consisted of large pieces of heavy material, such as wrought iron, iron from ship hulls or industrial machinery. No. 2 grades varied but were usually bundled collections of lighter metal. Collections from peddlers and scavengers were usually of No. 2 grade, as they could carry or haul a limited weight of material. As with rags, No. 1 grade old metal was the most desired and expensive classification; the large pieces contained fewer impurities or risks of varied alloy materials.

As was the case with rags, the classifications of No. 1 and No. 2 old metals were generally understood rather than formalized. Such understandings were not absolute; in 1872, for example, a New York firm imported 170 tons of wrought scrap iron from Europe for a price of sixty dollars per ton of 2,240 pounds No. 1 iron and thirty dollars per ton of No. 2 iron. Questions as to exactly what proportion of the shipments could actually be called No. 1 iron led to a dispute that ended up in court.[61] Disputes over material classification led to charges of fraud and several civil cases between customers and suppliers in New York, New Jersey, and Pennsylvania between 1860 and 1889.[62]

A Nuisance to Be Regulated

Though the demand for scrap in industrial America grew during the nineteenth century, the status of those working with the discarded materials did not. Rag and bone men had suffered low status in Europe due to fears about disease and concerns about criminal behavior. Due to scrap's novelty as a valued commodity, and to the dangerous and dirty nature of the work, few individuals who had opportunities in other occupations chose to work in the waste trades. Those who did were poor, usually immigrants with few business contacts. Scavenging was a widespread practice among the urban poor of the mid-nineteenth century. Women and children foraged the waterfront and city streets for items that could be sold to junk shops, which in turn sold them to nearby mills and artisans for reuse.[63]

Conflicts arose as scrap dealers developed customer and supplier bases. Secondhand material trading aroused the ire of public officials for both dangers to public health and perceived links between secondhand goods, theft, and violent crime. As early as 1817, New York City enacted a statute to prohibit secondhand dealers from doing business with minors who might be induced to steal and sell goods to secondhand dealers.[64] Police harassment of scavengers and junk shops in New York City increased around 1850. Other municipalities, including Boston and Philadelphia, enacted restrictions on the activities of junk peddlers and dealers in the second half of the nineteenth century. Many of the restrictions on junk sales were attempts to curb theft of goods that might be resold in shops.[65]

The dirty and dangerous nature of the waste trades was evident to those outside of the trades. Municipal governments viewed scrap, rag, and junk dealers as nuisances because of the physical hazards they posed. Concern over health issues, especially relating to the ability of rags to catch fire and spread disease, led to restrictions on rag sales toward the end of the century. Fires in New York City's rag shops threatened firemen and neighbors in densely built areas. Laws designed to limit rag and junk trading because of their threats to public health and comfort comprised one aspect of the nuisance laws proliferating in American cities and states in the nineteenth century.[66] Nineteenth-century enforcement of nuisance laws varied from state to state, but by the end of the century, judges in New Jersey, New York, and Pennsylvania used the laws to curb industrial nuisances. Nineteenth-century nuisance case law related to junk and rag dealers focused on licensing peddlers with the intent to limit theft and harassment by peddlers.[67]

Licensing laws were upheld by state courts, but with some limitations. In 1881, the New Jersey Supreme Court found that the city of New Brunswick had used peddler licensing laws solely to raise revenue for the city. The court ruled that convictions of license violations made with the primary intent to raise revenue for the city were illegal. Most of the appearances junk dealers made in New York, New Jersey, and Pennsylvania courts in the nineteenth century had less to do with nuisance laws and more to do with allegations of fraud, receiving stolen property, and disputes over weight and quality of delivered materials.[68]

Commodifying Waste

Between the late eighteenth century and the late nineteenth century, growing concern over waste as inefficiency in American industry transformed what had been a local, small trade in old materials to a commodities market spanning the nation's industrializing cities and across the Atlantic Ocean. Postconsumer and postindustrial sources of old materials increased, available materials were found in dumps, demolition sites, auctions of old industrial materials deemed obsolete by shipyards and railroads, and, in small amounts, from peddlers going door-to-door trading new goods for old. The combination of increased demand and rising consumption after the Civil War led to new opportunities for entrepreneurs to collect discarded materials and sell them to manufacturers.

By the close of the nineteenth century, material reuse in the United States had undergone several important transformations. Reuse of old materials in the household began to decline amid concern over hygiene and affordable alternatives in mass-produced goods. As American households began to throw away more materials, American manufacturers began to seek out more discarded materials. Demand from manufacturing industries for scrap materials transformed a collection of local activities concerned with small collections for use by blacksmiths and artisans into systems of specialized collecting activities dealing materials in larger volumes. Rags and old iron were now commodities valued in a transatlantic trade between the East Coast and Europe.

This burgeoning trade grew and adopted some professional standardization, but volatility and uncertainty remained. Though trade publications such as *Iron Age* began to publish weekly prices of rags, scrap iron, and other materials in the late nineteenth century, volatility continued to make dealers' lives uncertain. Setting prices required individual judgment as well as market trends. A good dealer needed to negotiate prices in a market where demand shifted by the week. Most firms amounted to little more than one man, his business contacts, and his skills. Dealers and customers identified standards by numerical designations, but what exactly constituted No. 1 iron scrap and No. 2 iron scrap caused many disputes because buyers and sellers frequently lacked common standards. As volume of materials traded increased, disputes became more frequent and serious. Waste trading was no longer the domain of beggars but was evolving into a major commodities market.

The structure of the scrap trade at the close of the nineteenth century leads to one final point. The firms that populated the trade were small and lacked competition from large, established institutions. The opportunities created in the scrap trade allowed a new wave of entrepreneurs to take advantage of the growing demand for secondary materials. Their activities shaped material reuse patterns as the nation industrialized.

2 | New American Enterprises

As the United States celebrated its centennial in 1876, Sigmund Dringer must have reflected upon his remarkable fortune in his new country. Dringer was an Austrian Jew who immigrated to New Jersey at the age of thirty. In the ten years after he entered the country, he rose from driving a junk wagon for a Newark junkman named Max Boehm for ten dollars a week to establishing his own yard in Paterson. According to the *New York Times*, he then "won the confidence of the large mill-owners there, extended his business rapidly," and by April of 1876 boasted an inventory of "4,000 tons of scrap iron and 1,700 tons of car wheels, worth upward of $100,000 in all, being the largest stock in the country. He claimed to be the most extensive junk-dealer in the United States, and was in a fair way to control the market. And yet this man can neither read nor write."[1]

Like Francis Bannerman, Dringer was an immigrant who had achieved success buying and selling scrap materials. Unlike Bannerman, he was Jewish. Dringer's success was remarkable, but his status as a Jewish immigrant who ascended from the bottom rungs of the business to a position of elevated social mobility resembled the stories of thousands of scrap peddlers and dealers who entered the trade in the late nineteenth and early twentieth centuries. Why did so many immigrants enter the scrap trade? And why were so many of these immigrants Jews?

These questions are central to the history of the scrap industry. The industry points to its roots in immigrants working hard to build empires out of junk piles; criticisms and caricatures of scrap dealers focus on traits related to ethnicity and foreign identity. One cannot understand how American patterns of salvage and reuse evolved without considering how the trade became so closely associated with immigrants, particularly eastern European Jews and Italians. Why these newcomers were able to shape the trade relates to the timing of economic, technological, and social developments, as well as cultural beliefs among both natives and newcomers regarding waste, hygiene, and aspirations. The values and goals of the immigrants were crucial; so too were the values of the millions of Americans who did not enter the trade and who found it distasteful. Understanding how immigrants created an industry out of waste requires an understanding of the nation they entered, how it created wastes to mine, and why others already in the country opted not to profit from those wastes.

Prosperous and Healthy

Scrap trade experienced its most conspicuous growth during a period of mass migration to the United States and industrial expansion. Industrialization had two important social effects: greater wages and salaries led many Americans to change their attitudes concerning consumption and waste, and industrialization produced new economic opportunities for immigrants. Growing firms after the Civil War created a substantial middle class of managers, bureaucrats, and clerical workers. As the middle class grew, it developed an ever-expanding consumer culture in the late nineteenth and early twentieth centuries, with the rise of department stores and catalogs catering to an expanding mass market. New patterns of consumption and waste disposal reshaped the way Americans valued objects, and changing attitudes provided the context for a national scrap trade dominated by immigrants.[2]

One major change by the end of the nineteenth century is that Americans—particularly middle-class Americans—had more objects, purchasing more clothing, watches, prefabricated housing, and other goods from a variety of outlets, including department stores, supermarkets, and mail-order catalogs.[3]

Concern over dress sense, proper eating etiquette, housekeeping, and public behavior all led to a proliferation of prescriptive literature and consumer items at the turn of the century designed to set members of the middle class apart from those in the working classes. Such indicators of class status played upon insecurities in the emerging middle class as its members attempted to define themselves as upwardly mobile.[4]

Consumption was not the only way the middle class attempted to identify itself; attention to health and hygiene as espoused by public health professionals became an important part of the middle-class sensibility in the late nineteenth century. Hygiene gained importance after the Civil War as theories of disease transmission evolved from a concern over miasmic vapors to the effects of germs. Crowded living conditions, smoky air, open sewers, and unpurified water meant epidemics with mortality rates in the thousands each year were common features of nineteenth-century urban life. As the germ theory of disease transmission gained favor among public health professionals, the goal of germ control led to a series of public health measures, some of them technological, others cultural. City streets once combined the functions of playground and dump, where children played and household waste was dumped. Even before the automobile began to dominate use of the street, concern for public health developed systems whereby public or private entities washed the streets and took refuse to dispose of in dumps or incinerators. By 1930, it was common for urban Americans to dispose of human waste in toilets in their residences, which transported the waste to sewers. Cans of garbage were filled and taken away from residential neighborhoods on a weekly basis.[5]

The home, now free of the stench of garbage and sewage in the streets, had running water to clean the body, home, clothing, and dishes. A combination of technological and organizational measures made life in America more sanitary in 1920 than it had been in 1880. To be sure, many residences, including tenements, lacked these amenities in the early twentieth century, but by this time they were seen as a problem for lacking them. Federal mortgage policies enacted during the New Deal specified that new homes suitable for federally insured home loans had to have basic sewage, water, and power amenities, policies that would reshape American homes as federally funded battlegrounds against germs by 1950.[6]

The battle over germs was not simply a technological one. At the same time, a cultural war against filth elevated the expectations of hygiene in the United States. Waste—be it human waste, dirt, or postconsumer items—was a threat to order and proper living. The proper storage and disposal of waste materials was critical, and what was considered waste was a more comprehensive list than ever before. Faced with her children's clothing becoming worn and ragged in 1850, a woman would mend the clothes as long as they could hold out, and once the pieces were no longer fit for wearing, she might incorporate them into a quilt, stuff the furniture with them, or find some other way of reusing them around the house. If the household could use extra money, the woman might ultimately sell the old clothing to one of the many peddlers collecting for the paper mills.[7]

In 1920, a middle-class woman was more likely to throw the old clothes into the garbage and ensure that the garbage collectors picked it up and removed it from the household. The 1920 house was much cleaner than its 1880 counterpart, and the inhabitants were also expected to be much cleaner. The woman and her family regularly bathed and washed with soap, an item freely available in stores and advertised heavily in magazines and newspapers. Over the next ten years soap was also advertised heavily on radio stations, creating the episodic dramas known as soap operas. Toothpaste and toothbrushes kept families' mouths clean, and cleansers to wash dishes, clothing, counters, and floors were all widely available for purchase and expected to be used in respectable households. Interestingly, the 1880 woman would likely find her 1920 counterpart incredibly wasteful by discarding the potentially useful fabric, and the 1920 woman would find the 1880 woman a source of dirt and germs that would likely harm her and her children's health.[8]

Mass consumption made it possible for the 1920 household to dispose of more items more quickly. Growing salaries allowed middle-class consumers purchasing power, and department stores, mail order catalogs, and lines of credit allowed them to use this power to conveniently buy new goods. Even if a woman in 1920 was not wealthy, she could live in a house where the clothing and furniture was all purchased from stores, the bathroom had copious amounts of purchased soap, disposable razors, and paper towels, and a wastebasket occupied most rooms of the house. In addition, an automobile might be parked outside. The house in 1920 had more things in it than it did forty years before, and for its middle-class inhabitants, the desire to have those things clean and orderly was a driving force in the classification of old materials as wastes to be eliminated from the household.

And there were many, many more old materials on hand to be eliminated by 1920 than there had been in 1880. Americans became more affluent in the last quarter of the nineteenth century, enjoying a standard of living that continued to improve into the early twentieth century. With affluence came waste; more goods available meant more refuse. Solid waste disposal escalated. Between 1903 and 1907, Pittsburgh's garbage almost doubled, increasing from 47,000 tons to 82,498 tons. Other metropolitan areas saw their garbage increase to the point that many engineers, chemists, city officials, journalists, and sanitarians voiced concern over excessive waste being a problem. Cities began using incinerators to reduce the amount of garbage stored in dumps. New York City even hired scavengers to sort through dumps for salvageable material. The municipal impulse to salvage was limited; most refuse was unwanted by manufacturers, but substantial markets existed for metals and rags, with demand for paper, wood waste, rubber, and glass. Whether salvaged, incinerated, or buried, postconsumer wastes posed a challenge to sanitarians. Consumption at the turn of the century was on a much smaller scale than that of the throwaway society of the late twentieth century, but the trend of consumption and disposal was rising.[9]

Clean Americans

At the same time the middle class grew and developed its distinct cultural concerns with hygiene, the working classes in the United States grew, in large part because of mass migrations from southern and eastern Europe between 1870 and World War I. Political instability, economic want, and ethnic repression contributed to the migration, as did the lure of wages in the rapidly industrializing cities of the Northeast and Midwest. These newcomers arrived with varied cultural practices, often ones quite different from those of middle-class Americans. The decades of mass migration produced tensions over many differences (including language and hygiene), causing debates over what American identity is and should be and how the new immigrants fit into the national culture.

Native-born Americans expected themselves and others to keep clean as a matter of moral responsibility. Unclean people could spread cholera, typhoid, scarlet fever, or any number of contagious diseases, threatening the safety of all. A responsible individual had to bathe regularly, wear clean clothes, and keep a sanitary household. Settlement houses, including Jane Addams's Hull House in Chicago, attempted to teach the urban poor and recent immigrants to adopt the rising American standards of hygiene, conflating cleanliness with American identity.

Many new immigrants in this war against germs lacked the facilities to uphold hygienic standards as easily as people living in middle-class neighborhoods. Prior to World War II, running water, gas, and electricity had become mainstream features of the new suburbs, but tenements and aging housing stock in the cities frequently left residents with few means to stay clean. Europeans who had come from areas that saw improvements in water filtration and sewage in the late nineteenth century did not share Americans' unique aversion to waste materials.

The differences between native-born Americans and immigrants regarding waste and hygiene manifested in many ways as settlement house workers and public health officials attempted to reform urban immigrants' practices, from using soap to discouraging the keeping of livestock to practicing accepted methods of kitchen hygiene. Many efforts were successful; after all, immigrants had their own taboos and practices regarding wastes and several were adaptable to their new homes. Careful attention to food preparation and hygiene, for example, may have resulted in lower infant mortality rates among New York City's foreign-born Jews when compared to those of other urban residents in 1910.[10] Yet the cultural differences produced different attitudes regarding handling wastes as commodities—and though most wastes were taboo, some were also coveted.

The War against Waste

The dual definition of waste found industries fighting inefficiency as cities fought filth. Railroads were the complex corporations to attempt sophisticated managerial systems designed to maximize efficiency, but the ethos of efficiency made its way into every corporate boardroom by the early twentieth century. Meatpackers bragged about using "every part of the pig but the squeal" to reduce waste. Frederick W. Taylor became famous for his principles of scientific management, reshaping industries from the railroads to steel to automobile manufacturing. Waste in industry was a problem solved by engineering, efficient management, and technological innovation. The metals industries were particularly cognizant of the value of scrap materials, as not only were scrap metals much more affordable than virgin ore, but concerns over the ultimate abundance of nonrenewable commodities underscored the importance of reclaiming postconsumer and postindustrial supplies.[11] Many American industries, including steel and automobile manufacturers, have continued to seek out ways of reusing their own materials and acquiring discards from other sources since the late nineteenth century, establishing practices that continue today.[12]

Economic and technological dynamics produced new demands for scrap materials, especially in the period of heavy industrialization between 1870 and 1920. As the paper industry adopted wood pulp at the end of the nineteenth century, old iron replaced rags as the most widely traded scrap material. Changes in the steel industry bolstered demand for old iron. Under the lead of Carnegie Steel at the end of the nineteenth century, and after 1901 the industrial giant U.S. Steel, American steelmakers expanded production throughout the first half of the century. As the steel industry grew, its preferred process shifted from the Bessemer process to the open-hearth process, which employed much greater amounts of scrap iron. Andrew Carnegie shrewdly monopolized the vast ores of high-phosphorus iron in Minnesota in the 1890s, forcing his competitors to adopt alternate methods of production in order to stay competitive. Building open-hearth furnaces and using scrap iron and steel was an increasingly popular option. By 1910, it was a necessity. The Gary Works built several open-hearth furnaces between 1908 and 1920, and many of U.S. Steel's smaller competitors switched over from Bessemer furnaces to open-hearth furnaces in order to survive. The transition to open-hearth steel combined with the expanding markets for steel to trigger new demands for scrap iron and steel. At the same time, steel production expanded as the industry supplied the materials that made houses, skyscrapers, ships, railroads, armor, industrial machinery, automobiles, and other goods.

The growth of steel catalyzed industrial development in the new cities of the Midwest. Pittsburgh enjoyed its largest surge of growth during this period, as the vast resources of bituminous coal in western Pennsylvania and West Virginia provided fuel for the growing industry. Chicago's status as the gateway to the West, connecting both coasts through its network of rail lines, allowed the city to withstand two serious fires in the 1870s and increase both its industrial capacity and population in every decade between the Civil War and World War I. Just to the southeast of the city along Lake Michigan, U.S. Steel's Gary Works was the largest steelworks in the world when it opened in 1907. By 1920, Detroit, Flint, Cadillac, and several smaller Michigan towns were destinations for migrants seeking work in the automotive industry. Steel and the things that could be made from steel fueled the growth of the urban Midwest, producing thousands of jobs and reshaping communities.

New Opportunities

As the United States industrialized, it attracted millions of immigrants from Europe seeking greater economic opportunity. Over 33 million people immigrated to the United States between 1820 and the early 1920s. Between 1820 and 1880, over 85 percent of all immigrants came from the United Kingdom, Germany, and Scandinavia. After 1880, the number of immigrants from Italy, eastern Europe, and the Austro-Hungarian Empire rose, accounting for more than half of all immigration by 1896. Jews fleeing poverty, religious persecution, and overcrowding in the ghettos of the Jewish Pale represented a significant fraction of the new immigration, transforming American Jewry from a population dominated by immigrants from Western Europe to one where most of the population originated in the Pale. Between 1880 and 1920, an average of about six million people arrived in the United States in each decade. From 1900 through 1909, more than eight million newcomers entered the United States, settling primarily in northeastern industrial cities.[13]

The newcomers shaped the labor force of numerous industries—for example, the steel mills of Pittsburgh were quickly staffed by thousands of Slavic laborers. Italian-born individuals helped shape the construction industry in several cities. Pittsburgh's population nearly doubled between 1880 and 1890 as the steel industry grew. Pittsburgh's large industrial concerns grew; in 1899, thirty-two Pittsburgh firms employed between one thousand and ten thousand workers and four firms employed over ten thousand workers. Carnegie Steel, the city's largest company, had over twenty-three thousand employees. The new jobs thought to be available for workers lacking formal skills from this industrial development attracted newcomers from southern and eastern Europe, as well as African Americans from the South.[14]

Mass immigration resulted from a combination of economic and social pressures in Europe and opportunities emerging in the United States due to industrial expansion. The archetypal stories of immigrant life and work in the United States during this period revolve around factories. The Slavs in the steel mills of *Out of This Furnace*, the

Lithuanian meatpackers of *The Jungle*, and the young Jewish girls killed in the Triangle Factory fire portray the immigrant as industrial employee, usually in a menial position defined as unskilled. Even if we suppose that identifying such work as unskilled obscures the tangible skills and abilities required to do grueling work, this portrait of the immigrant work experience is still incomplete, for thousands of migrants came to the United States not to be someone's employee but to run their own businesses.[15]

Immigrants who did not or, due to discrimination, could not seek work in the large mills found opportunities to advance in new enterprises related to the scavenging many immigrants already did in order to survive. Jewish immigrants, for example, established niches in the garment and grocery trades, and many established businesses as peddlers selling food, clothing, and other small goods in both urban and rural areas. Scrap emerged as an option for those who wished to start their own businesses. The evolving "waste trades," comprising scrap metal trading, rag trading, scavenging, and garbage hauling, offered opportunities to urban newcomers. These trades in the late nineteenth and early twentieth centuries had what sociologist Roger Waldinger calls an opportunity structure conducive to high levels of immigrant business ownership. In his study of the garment trades in New York City, Waldinger identified several characteristics within industries that featured large proportions of entrepreneurship from foreign-born individuals. Two of the most important characteristics were little competition from established natives and low required investment costs.[16]

The scrap trade in the late nineteenth century featured few native-born Americans and few economic barriers to entry. Association with filth precluded many natives with other prospects from entering the waste trades, for identifying and collecting usable scrap was not a simple or pleasant task. Successful scrap collection involved sorting through piles of refuse and having the ability to discern valuable material. The work required uncomfortable physical labor in city dumps, mills' discard piles, and other settings considered unhealthful and unsanitary. The materials posed risks to the collectors. Rags were flammable and could harbor diseases. Junk-heaps could contain jagged objects, poisonous materials, and any number of hazards to the people working in them. The work, by its nature, was the kind of low-status, high-risk work sociologist Stuart E. Perry called "dirty work."[17] Since waste materials were considered unsanitary and contact with them

Figure 2. Street peddler on the Lower East Side, New York City. Peddling scrap metal and rags provided thousands of immigrants opportunities to establish their own businesses. *Courtesy Picture Collection, Branch Libraries, New York Public Library, Astor, Lenox and Tilden Foundations.*

exposed the handler to disease and injury, few individuals who had opportunities in other occupations chose to work in the waste trades. Those who did were poor, usually immigrants with few business contacts. Immigrant women and children scavenged for scraps in mid-nineteenth century cities for subsistence, and immigrant men scavenged and peddled scrap materials.[18]

America's industrializing cities provided many sources for scavengers. City dumps replaced streets as centers for refuse, allowing scavengers to sort postconsumer waste at centralized locations.[19] Construction and demolition sites featured portable amounts of iron, copper, lead, wood, and other materials. Factories, railroads, and mills produced waste in the form of broken or worn-out equipment and unused materials. They also were sites where less scrupulous scavengers could steal small amounts of scrap in the form of old rails, small fixtures, and unguarded machinery. Enterprising peddlers traded new consumer products for obsolete household items. The urban poor had opportunities to generate income from scavenging and selling their collections to junk shops or mills.[20]

The rural hinterlands beyond city centers provided opportunities to reclaim old materials. I. H. Schlezinger was a Jewish immigrant from Austria-Hungary who came to Columbus at the turn of the twentieth century. His son Edward recalled that I.H. worked as a peddler, going out in the country with his horse and wagon to visit the area farmers. "The women would give him orders to bring out to them things—to pick up for them—and then he would trade—trading—it was called—then if they accumulated—like the husbands had a pile of old iron on the farm—scrap iron—he would take that in return."[21]

I. H. Schlezinger's business indicated the growing industrialization of the nation produced new systems of trade between the city and countryside as demand from the urban mills led to forays to collect rural scrap. Immigrants who did not wish to work in factories found that the demand of those factories for more material allowed them to have careers supplying factories from whatever they could collect from urban neighborhoods or the farms outside the city. Industrialization's effects on the workforce was not limited to life in the factories but extended to the creation of small businesses that benefited (and helped shape) mass production.

Historical accounts of American cities in the nineteenth century suggest heavy participation by immigrants in scavenging, junk collecting, and scrap dealing in the late nineteenth and early twentieth centuries, and demographic data indicates that this suggestion is accurate. A correlation between place of birth and occupation in the Integrated Public Use Microdata Series (IPUMS) sample of the United States Census of Population for 1880 indicates that over 70 percent of the workers in the waste trades in 1880 were born in Europe. The typical waste trade worker in 1880 was male, married, the head of the household, and born in Europe, with the largest groups coming from Germany, Ireland, and Poland. While narrative accounts suggest women continued to scavenge, they did not enumerate "waste trade" as their occupation, indicating a limit in enumerators' questions that may have accentuated the gendered nature of the work. It is also possible that individuals who scavenged to augment household incomes did not identify themselves as scavengers to the census.[22]

The census data identifying work with waste materials as an almost uniformly male activity conflicts with anecdotal accounts of female scavengers yet also reflects a growing formalization and gendering of the industry. While men and women alike had scavenged dumps, streets, demolition sites, and the waterfront, the census did not enumerate women as being employed in the waste trades, possibly due to bias by enumerators to identify individuals working in scrap yards as workers and not scavengers.

The census's methods also revealed a very white workforce—if we consider whiteness as defined by the census with caveats over how "white" native-born Americans viewed Jews, Italians, and Slavs between 1880 and 1920. Most of the men enumerated were white (93 percent), as opposed to black or Chinese. As with the data indicating waste trade workers were mostly male, the conclusions defining a white workforce may have been a function of the method used to enumerate workers. Stories of African Americans working as scavengers in northern cities abound: African Americans worked as yard labor in urban scrap yards and ran secondhand goods and junk businesses in the South. These individuals, however, may have been missed by the census. Transient scavengers (be they white, black, male, or female) likely were not listed as waste trade workers, if they were even recorded for the census. And while African American junk-shop owners may have had stable businesses, they may not have tapped into the industrial network feeding large mills, perhaps causing census takers to not consider them with the larger workforce. Finally, even individuals working in yards that were in that industrial network were often people who worked multiple jobs or were employed seasonally and may not have described themselves primarily as waste trade workers to the census takers. The white male image of the waste trades as developed by census information is overwhelming, but it may be obscuring a more varied portrait of the business of reclamation and waste management in the late nineteenth century.

Keeping reservations about biases involving race and gender in mind, the census provides a portrait in keeping with the idea that the waste trades offered opportunities for European immigrant men who wished to start their own businesses. A large proportion of the individuals born in Germany and Poland were likely Jewish, though the 1880 census does not feature a positive indicator of Jewish status. The presence of established Jewish scrap dealers such as Sigmund Dringer, and Cline and Bernstein, of Nashville suggests Jews had already achieved some success in the business prior to 1880. By 1920, so many Jews—many fleeing repression in Russia and the Ukraine—had started businesses trading in scrap iron, rags, and other secondary materials that the public face of the scrap dealer was the face of a Jew.

The 1920 census allows for a more precise measure of Jewish identity than does the 1880 census. As Susan Cotts Watkins observed in her work on the 1910 census, the presence of native tongue of both the enumerated individual and the individual's parents as a category allows for the most precise measure of Jewish identity in census records. As Jews were the only group likely to identify Yiddish or Hebrew as a native tongue, identifying Yiddish-speaking individuals, and their children, as Jews is a reasonable inference.[23]

Sixty-eight percent of the junk workers in the 1920 IPUMS were from countries in eastern and central Europe, where Jews had emigrated from in large numbers over the previous four decades. A majority of the junk workers from eastern Europe spoke a native tongue of Yiddish, Hebrew, or "Jewish." Even leaving aside the caveat that this technique may omit Jewish individuals whose mother tongue may have been recorded as Russian, German, or another language, defining workers' ethnicity by native tongue indicates that first-generation Jewish immigrants were by far the most represented group in America's junk trade in 1920.[24] Contemporary observations agree with the statistics. By the mid-1930s, the editors of *Fortune* estimated that the scrap metal industry was 90 percent Jewish owned.[25]

The large Jewish and Italian representation in the scrap metal and rag trades was in part due to the mass waves of immigration from eastern Europe coinciding with the growth of demand for scrap materials by American industry; certainly the timing provided easy entry into a trade where demand was high, initial investment costs were low, and an individual who could assess good material from worthless material and offer the goods to an industrial customer could, as Sigmund Dringer did, enjoy rapid upward economic mobility.

Yet timing alone cannot be considered the reason for immigrants' interest in scrap iron and steel. The opportunities the new industry offered matched the goals of many Jewish immigrants. Jews had a long tradition of entrepreneurial activities from peddling to shop keeping; Jews and Italians both started retail businesses with pushcarts and developed small grocery stores. For Jews, the impulse to start a business was due in part to the advantages of not having a supervisor who could discriminate against their religious practices, fire at will, or refuse to hire. As historian Alan Kraut argues, even if a shopkeeper's or peddler's income was lower than a factory worker's, the lure to be one's own boss was very strong. A Jew in business for himself could set his own schedule and priorities without worrying about whether a Christian employer would compel him to work on a Friday evening or discriminate against him because of his religion. Such concerns were not abstract to eastern European Jews who had worked for gentiles in their native lands, and at a time when religious persecution was on the rise, the opportunities for freedom in the United States extended to the promise of being one's own boss.[26]

Jews also peddled produce and other goods and operated a variety of shops. The scrap trade was similar to the produce trade. Both trades required little initial capital investment yet required evaluation of inventories suitable for sale, as well as the ability to network with other buyers and sellers. Historian Oliver Pollak notes that the retail peddler trade was on the wane by the turn of the century as competition from established shops led to a loss of customers. Junk peddlers benefited from mass consumerism and planned obsolescence, and junk peddling became a growth occupation.[27]

Jewish participation in New York City's secondhand goods trade was widespread by 1880, including scrap yards, junk shops, and pawnbrokers. A New York Times report in 1866 on the trade in stolen goods observed that Jews had "a monopoly of the pawnbroking business in this City, and many of them have acquired great wealth nearby. They are sharp and shrewd, always driving a hard bargain." A combination of natural talents and constant practice, the Times argued, "made them as good judges of human nature as they are of the value of an article. They can tell at a glance whether or not their customer is a constant visitor at the pawnbroker's counter, and woe be to the individual whose manner shows him or her to be a novice in such transactions."[28] The Times recognized the attributes of a successful pawnbroker as skills, though not ones to be admired. Dealers in secondhand goods were not respectable citizens, but "low, degraded people, of Irish origin generally, although many who speak the German language are engaged in the business." The "degraded people" reflected both xenophobia and middle-class disgust of the trade. At the same time the succession of Irish dealers by German-speaking dealers reflected a shift in the countries of origin of waste-trade workers as immigration patterns changed between 1880 and 1920.[29]

As the Times report made clear, Jews were one of several immigrant groups who participated widely in the waste trades after 1880.

The 1920 IPUMS sample indicates that the largest groups working in junk were first-generation immigrants whose native languages were either Yiddish or Italian. Yiddish-speaking, foreign-born individuals were represented in numbers over twenty times their representation in the general population. Italian-born individuals were represented in numbers about twelve times their representation in the general population.

Italian immigrants participated in New York City's rag dealing and garbage sorting in large numbers. In 1881, the *Times* reported that the people who "separate from valueless material the atoms that can be put to use again are almost entirely Italians . . . as industrious as ants, and, apparently, have eyes for nothing but the bits and particles that go to fill their bags."[30]

The *Times* report was perhaps biased because the reporter filed his story from the heavily Italian Five Points neighborhood of the city, but New York City street-cleaning commissioner Col. George E. Waring Jr. employed Italian immigrants in the 1890s as scow trimmers. Waring reasoned that Italians were "a race with a genius for rag-and-bone picking and for subsisting on rejected trifles of food."[31] Waring's stereotyping reflected a conflation of subcitizen identity with handling waste; his comments reflect both a cultural bias and the degree to which Italians participated in the waste trades in the 1890s.

As Jewish and Italian immigrants continued to populate cities in the Northeast, the number of scrap firms multiplied. These businesses were usually very small, requiring modest capital to get started. Herman Landeson of the Landeson Metals Company recalled that he entered the scrap business "in 1913 when Samuel Sher and I started as partners, with the usual equipment of those days, horse and wagon, peddling through machine shops and garages. The business then consisted mostly of buying tires, tubes, and an occasional radiator, aluminum crank case, or battery box."[32]

Landeson's story was not unique; the scrap industry's history between 1880 and 1920 is rife with stories of young immigrants starting out with a sack of bottles, a pushcart full of scrap metal, and a small list of local customers.[33] The number of scrap businesses in large mercantile cities on the East Coast multiplied between 1850 and 1900. Philadelphia's business directories indicate that the junk, rag, and scrap businesses grew significantly over the last half of the nineteenth century. After a period of steady growth between 1850 and 1880, the number of junk, rag, and scrap establishments in the city grew to about

300 by 1880, then leveled off until 1910, when the increase in iron and steel scrap dealers led to the total number of firms almost doubling (529 firms in 1910).[34] As industrial production in the Midwest increased, the scrap industry followed. Detroit's scrap trade grew from sixty firms in 1890 to 127 in 1910 and 296 in 1920.[35] Chicago's scrap firms increased their numbers from 140 in 1890 to 290 in 1910 and 471 in 1917.[36]

Scrap firms proliferated in the city centers. Recent immigrants frequently settled in inner-city areas such as Harlem in New York City, the near West Side in Chicago, and the Hill District in Pittsburgh. These areas were within walking distance to large employers; they also were proximate to loud, dirty industrial practices that had begun to lead upper-class residents to search for quieter, cleaner residences on the urban periphery. Mass suburbanization would not occur until after World War II, but the dynamics leading to class segregation with the lower classes concentrating near the central business district were in place by 1880. Scrap firms joined in the industrial nuisance generated in growing American cities. Many businesses operated out of the yard or house of the proprietor, producing sounds, odors, and sights that neighbors might find objectionable.

The character of the nuisance depended upon the type of work done by individuals in the trade. This work was increasingly specialized as the trade moved beyond subsistence scavenging by randomly assorted individuals to a loosely organized network canvassing cities and servicing regular customers. The nineteenth-century scrap trade involved individuals working on at least three levels: collectors, yard or shop owners who purchased goods from collectors and sold them to local industry, and the occasional rag dealer who bought rags from Europe and sold them to American paper mills. By the turn of the century, the scrap industry's increased volume reshaped the industry so that workers operated at five levels of activity.

Collectors included individuals who gathered small amounts of light materials by scavenging. These people included women and children who sold their collections for subsistence to peddlers, dealers, or processors. Collectors were likely excluded from the census as waste trade workers, but they remained a significant presence in American cities in the early twentieth century.

Peddlers differed from collectors in that they had some equipment and were typically adult males, as were the individuals at the rest of the levels of the industry. Peddlers bought and sold small amounts of materials, working with sacks, pushcarts, and sometimes horses and buggies. Initially, peddlers traded retail goods in exchange for postconsumer materials, including rags, bottles, and broken pots, then they shifted to exchanging cash for postconsumer scrap.

Dealers differed from peddlers in that they operated from a fixed location, usually a yard or shops where they stored their materials. Dealers traded postconsumer scrap and purchased industrial scrap, military surplus, and collections from scavengers, peddlers, and other dealers, trading larger volumes of materials than peddlers did.

Processors were dealers who also processed materials and invested in technology to facilitate processing. Scrap metal dealers used shears, hammers, and torches to cut and shape scrap into portable shapes and sizes. Rag and rubber processors sorted their materials manually with minimal use of technology.

Brokers occupied the highest level in the waste trades. They became the most common interface between the waste trades and other industries because brokers took orders from industrial customers and then purchased materials from several dealers and processors to fill the orders. Brokers differed from dealers in scale; like dealers, many brokers also processed materials. During the nineteenth century, brokers engaged in the transatlantic rag trade filling large orders for paper mills; after the turn of the century, iron and steel brokers became important conduits between dealers and large steel firms.

Though the census data do not correspond to these classifications, one could assume that women and children made up a larger proportion of workers at the level of collector/scavenger than any other level of the industry. The remaining four categories presuppose some level of investment, business contacts, and formal business practices that were more readily available to men in the nineteenth century.

The customers of these businesses were fellow dealers and larger industrial concerns primarily owned by white men. Racial barriers may have precluded African American and Asian Americans from engaging in much business with customers, keeping their numbers as dealers, brokers, and processors low. Despite ethnic differences that could cause conflict, white, European-born males managed to engage in business with large industrial customers and participate at all levels of the evolving scrap industry.

Self-Made Men

The immigrants who opted to enter the trade at the turn of the century had the opportunity to experience significant and rapid mobility. The rise of steel allowed scrap companies to grow from one-man peddling operations to scrap dealing and processing yards investing in processing technologies and larger labor forces. Sigmund Dringer's success in the 1870s was an early but not unique example. Julius Solomon rose from his first days in Pittsburgh as a thirteen-year-old peddler from Poland in 1863 to run a scrap iron business that made him a wealthy man. He passed away in 1911, leaving a scrap iron, salvage, and used machinery business with locations in Pittsburgh and Beaver Falls, Pennsylvania, to his four sons. By the time of his death, he had risen from peddler to philanthropist and was identified by Pittsburgh historian George Fleming as "the self-made man who had reached a high position in the world but still gave lavishly of his personal interest and of his money to help those in need."[37]

One of the largest of all scrap firms began as a small family operation run by an immigrant Jew. Hirsch Luria emigrated from Germany in the early 1890s. Upon his arrival in Reading, Pennsylvania, he became a junk peddler. Displaying a talent for evaluating desirable scrap, he parlayed his peddling income into a scrap yard, specializing in scrap iron and steel. The yard was successful, and twenty years after he started the business, Hirsch passed it on to his sons Alex and Max. The sons expanded Luria Brothers, opening scrap yards in Philadelphia, Pittsburgh, Cleveland, and other cities, growing their father's business to become one of the major scrap iron and steel firms in the world. They also established an import/export business that prospered after World War I.[38]

Labor in these new, small businesses often involved relatives. Luria Brothers drafted several members of the Luria family into service in the yards in the early years, and many firms employed their children once they became old enough to handle tools in the yard. As an adult, Louis Galamba recalled that "Father opened a waste material business [in Webb City, Missouri, in 1903] and in 1905 I assisted him after school hours and worked during vacation periods." The business thrived and Galamba grew up to be president of the international trading firm Sonken-Galamba half a century after his after-school labors.[39]

Drafting family members was both a cost-effective form of employment for the small business owner and a way to provide opportunities to the younger generation. Milton K. Mahler reminisced in 1980 that the early-twentieth-century scrap business grew as families grew, starting as one-man operations with a horse and wagon. According to Mahler, when a son comes along, "they look for a coal yard that went out of business. And the reason he looks for a coal yard that went out of business is there is a scale and a railroad site and a little office. Another son comes along, he has to expand. A son-in-law comes along, he has to expand."[40]

Decisions concerning firm growth depended upon both market forces and family structure. Though hard decisions concerning the direction of the business were common, an individual attempting to start a firm could do so without worrying too much about financing. Immigrants could enter the scrap industry with minimal investments in technology. Scavengers and peddlers could function with little more than a sack. Peddlers often invested in pushcarts or rented horses and buggies in order to increase the amount of materials they could collect. The low level of capital needed to start in the industry was one condition that allowed immigrants opportunities to start scrap businesses. Between 1880 and 1920, technological developments that made processing scrap materials more efficient raised the cost of doing business over simply working with one's own hands and a sack, but investment costs remained modest throughout the period. Most of the labor involved manual sorting. Scrap dealers used sledgehammers to break apart large pieces of scrap metal, work that went quicker in subzero temperatures, as iron and steel were more brittle when frozen.[41]

Hirsch Luria's yard was one of many that used alligator shears to cut metal into smaller, more manageable pieces. Alligator shears were first used in scrap yards around 1880. One used an alligator shear by placing the metal between its jaws and bringing the top jaw down. Scrap shears could cost fifty dollars to more than one thousand dollars, depending on the size and capacity of the shear. Manufacturers made a wide variety of alligator scrap shears, with some catalogs boasting over twenty varieties. The Canton Foundry and Machine Company of Milwaukee, for example, advertised five sizes of scrap shears in 1920.[42]

Lifting magnets attached to cranes entered steel mills in 1898; the first lifting magnet used for handling scrap won a gold medal at the 1904 World's Fair in Saint Louis. By 1906, lifting magnets were common in scrap yards, as were gas cutting torches. Yard workers used torches to cut large pieces of scrap such as ship hulls that would not fit in shears. Scrap men purchased blowtorch gas, which came in riveted tanks about the size of a galvanized hot water boiler, allowing faster processing of heavy scrap iron (the most coveted grade of scrap iron) than sledgehammers could provide.[43] Balers had widespread agricultural uses at the turn of the century. By 1920, scrap dealers realized that balers used to sort and store hay could bundle light scrap into cubes of dense material more conducive for easy storage and transportation. With a baler, a scrap processor could take what was typically a lower-value material and produce heavier stocks that took up less space for the weight in railroad gondola cars. As World War I brought new revenues into larger yards, some firms invested in magnetic cranes and planned yard layouts to speed processing in the yards.[44]

The technology employed in scrap yards increased the hazards of an already dangerous industry. Workmen's compensation lawsuits in the 1910s and 1920s revealed that the new technologies produced new dangers. Cranes tipped over. Alligator shears could speed processing of metal, but they also resulted in the loss of fingers, hands, and arms in shear mishaps. Gas torches elevated the possibility of burns and explosions in yards. On 5 December 1914, Morris Brenner was working with a torch in the Reliable Junk Company's yard in Baltimore when an explosion killed him, a story that was unfortunately not rare in scrap yards. Firm owners and yard labor alike were at risk of death or dismemberment from the new machinery they used every day.[45]

The new technologies elevated danger for waste trade workers, but they also allowed the businesses to sort and process materials rapidly, allowing them to fill larger orders from customers who hungered for more material. The investments in larger yards and more sophisticated technology indicated that scrap firms were playing more prominent roles in American industrial life, extending the use of old materials and making industrial activity more efficient and profitable for scrap supplier and buyer alike. Increased danger was one complication of the newfound opportunities, one that most dealers opted to chance.[46]

At the end of World War I, the United States had experienced half a century of heavy industrialization and mass immigration. Both factors transformed the waste trades to the extent that by 1920 dealers in rags, old metals, and other secondary materials had established a multimillion dollar industry operated in large part by men whose families had not lived in the United States forty years earlier. Demand for scrap iron during World War I made ferrous metal—already becoming a standard material for industrial production—even more valuable, expanding the market and making some dealers—including David J. Joseph in Cincinnati and Hirsch Luria's sons working as Luria Brothers in Reading—very wealthy men.

Building an Industry

American attitudes concerning waste joined industrial desires for affordable materials in providing the possibility of an immigrant-dominated industry at the turn of the twentieth century. Immigrants from southern and eastern Europe took advantage of that possibility because it fulfilled specific goals. As demand for scrap materials by manufacturers grew, unease with waste among the growing American middle class also grew. At the same time, millions of eastern and southern European immigrants entered the United States in search of economic independence. Industrial demand and cultural unease concerning wastes combined to produce opportunities for thousands of these immigrants collecting, sorting, processing, and selling scrap materials. Many Jewish newcomers preferred to start their own businesses to avoid discrimination and achieve economic independence, and the growing number of Jewish tailors, grocers, and peddlers during this period indicates the range of trades the budding new entrepreneurs explored as they sought to fulfill their aspirations.

The demand for scrap materials allowed hundreds of new scrap firms to develop in urban America. Lack of established firms and low investment costs allowed newcomers opportunities to establish their businesses and subsequently build upon initial successes. Handling scrap materials held little attraction to most natives, as scrap lacked proven profit-generating ability and involved work in unsanitary conditions. Those willing to handle waste materials in order to own their own businesses enjoyed an opportunity structure that allowed not only easy entry for newcomers but also quick and substantial growth of firms.

Starting a scrap firm required very little start-up capital. Other than a sack, pushcart, or horse and buggy to collect goods, all an individual needed was the ability and desire to sort and sell scrap. For thousands of immigrants, particularly Jews fleeing religious persecution and wary of discrimination in the American workplace, handling waste materials was a very attractive prospect.[47]

Innovative immigrants took these small one-person businesses and created a loosely organized network of firms reclaiming scrap materials from postconsumer and postindustrial sources and returning them to manufacturers, completing the cycle of production. The most successful of these scrap dealers, such as Sigmund Dringer and the Luria Brothers, were able to expand their businesses, purchasing land and processing technology to handle greater volumes of material. By 1920, immigrants had created a trade with firms ranging from one man with a sack to businesses employing dozens of employees handling tons of steel for industrial customers.

Economic downturns such as the Depression following World War I were reminders that though scrap trading promised opportunities, it was also quite volatile. Many small firms failed due to lack of demand, work-related injuries to operators, and other mishaps. Opportunities for new entrepreneurs were great, and immigrants and their families could enjoy upward mobility from peddler to dealer to broker in the space of a few short years.[48]

Handling scrap presented opportunities to immigrants; it also presented dangers. The unpleasant associations with pollution ensured little competition from natives and, conversely, ensured scorn from natives. By World War I, xenophobia and hygienic concerns produced an image of a stereotypical hook-nosed Jewish junk peddler or swarthy Italian scavenger bent on dirtying the streets and morals of urban America. The caricatures represented thousands of new entrepreneurs who were conspicuous in cities in the Northeast and Midwest, transgressing taboos about waste (as pollution) to profit from concerns about waste (as inefficiency). Transgressions resulted in conflicts with the dominant American attitudes about waste and hygiene that the new businesses were forced to address after the turn of the century.

3 | Nuisance or Necessity?

Abe and Sam Levinson were successes. Second-generation Jewish immigrants living in Pittsburgh, the two brothers ran a profitable scrap metal yard and rag shop in the city's Hill District, close to downtown. Their father, James Levinson, had founded the business on Pride Street, starting with a horse and wagon. His grandson Aaron remembers the yard had "all the junk that he could find—old tires and old brass bedsteads, pots and pans, metal of any kind."[1]

James passed away in 1917, leaving the business to his sons, who continued to build on their father's achievements, making the business even more profitable. Within a decade of taking over the family business, however, Abe and Sam had stopped buying and selling scrap iron and rags, opting to close the Pride Street businesses in 1926. Instead, they changed the direction of the firm, opening a new steel fabrication plant at Twenty-third and Josephine Streets on Pittsburgh's South Side. Even though Levinson's junk business was very successful, within a few short years of taking over the family business, Abe and Sam abandoned the scrap trade and founded Levinson steel. Why would successful scrap dealers leave the trade as the market for secondary materials was growing?[2]

James, Abe, and Sam Levinson all enjoyed the opportunities offered by the scrap trade in the early twentieth century. The first two decades of the century marked an extraordinary period of growth for scrap dealers as large industry began demanding mass quantities of scrap iron. By 1917, *Scientific American* reported that the annual business in scrap iron increased from one hundred million to a billion dollars.[3] Between 1917 and 1929, demand for iron and steel would further expand the market for scrap iron. The mass consumption of the 1920s brought obsolescence based on style rather than technological function. General Motors pioneered the concept of the annual model in automobile sales, and manufacturers of washing machines, gas ranges, and other durable goods added stylistic considerations to the marketing and design of goods whose useful lives would otherwise span many years.[4]

A consumption ethic based on style rather than functionality spurred disposal of durable goods. Thousands of junk shops and yards welcomed millions of mass-produced metal goods that could be resold as secondhand appliances or processed as scrap. In 1900, 5.1 million gross tons of ferrous scrap were sold on the market, mostly to domestic sources; eight years later that figure rose to 8.6 million gross tons, and consumption in World War I led to the sale of 12.7 billion gross tons in 1917.[5] The automobile graveyard became a new, specialized junkyard where customers could purchase obsolete automobiles for scrap or purchase individual parts off of junked automobiles in order to repair other automobiles. In 1920, Detroit's business directory featured listings for thirty-six automobile salvage yards, and by the end of the decade, automobile graveyards were common in American cities by the end of the 1920s. Chicago's scrap businesses increased from 140 to 417 between 1890 and 1917.[6]

By the end of World War I, the market for scrap iron and steel had made wealthy men of many dealers. The collection of small firms built by immigrants over the past four decades had grown into a flexible network of brokers, dealers, peddlers, and scavengers, with the largest firms enjoying international trade and relationships with the nation's largest industrial producers. In 1918, journalist George Manlove exclaimed that the iron and steel scrap industry "has risen from nothing to a position of full dignity and is just now coming into its own, a recognized member of the iron and steel family."[7]

The Crookedest of Any Business

While Manlove was correct about the industry's rapid rise, his assertion that it had risen to a position of full dignity was off the mark. The hundreds of new scrap dealers competing to sell their goods to a smaller number of industrial customers encountered conflicts relating to their business practices and handling of waste materials. Inherent in the dialogue between the scrap industry and its critics were different aspects of environmental values and concepts of legitimacy prevalent in American thought between 1880 and 1930, conflicts that pitted an established business class of natives against upstart immigrants.

One reason immigrants were able to get such a strong foothold in scrap was the trade's reputation as dirty, unscrupulous, and dangerous; the unpleasant aspects of the work were a factor in many earlier generations of immigrant waste trade workers searching for alternative occupations—once they established the social and economic means. As junk shops, scrap yards, and peddlers proliferated, customers, neighbors, and police increasingly saw them as centers of physical and moral ills.

New scrap dealers faced several problems succeeding in their chosen field. Handling scrap materials was hazardous work, exposing the handler to tetanus, injury, and even disease and fire, if rags were a handled commodity. In order to succeed in business, scrap dealers had to overcome the dangers posed by the work. They also needed to establish relationships with customers in order to get repeat business in a field populated by many small independent firms and relatively few customers. Dealers competed to offer the best quality materials at the lowest prices. Some also found unscrupulous ways to undercut the competition, offering materials they could not deliver or misrepresenting the quality of their materials.

A common complaint involved the quality of materials sold. By 1860, buyers and sellers of old rags, iron, and other materials developed common understandings of material classifications, with No. 1 material being more desirable than No. 2. No written agreements defining the classifications existed, and many disputes centered on the question of material quality, leading to lawsuits and accusations dating back to the Civil War.[8]

Sigmund Dringer was a remarkable success in his first decade as a scrap iron dealer in New Jersey, but by 1876 he had gotten into considerable trouble with one of his largest customers. That year, the Erie Railroad's Receiver, Hugh J. Jewett, charged in New Jersey's Court of Chancery in Trenton that Dringer had induced Col. Henry Bowman, the purchasing agent for the Erie Railway, to sell Dringer all the old iron and other metal accumulating on the road at prices well below

the market value of the metal. Jewett accused the scrap dealer of conspiring with Bowman and filed suit against both. He also charged many
other officials and workers of the railway with conspiring with Dringer,
alleging "a vast system of corruption, extending even to the common
laborers in the company's yards."[9]

Jewett's suit alleged that Dringer was a skillfully manipulative operator who had convinced the Erie Railroad's employees to permit the
dealer to carry off immense quantities of No. 1 scrap iron (worth eighteen to twenty dollars per ton) charged only as No. 2 scrap (worth ten
to twelve dollars per ton). According to this plot, Dringer loaded cars
with No. 2 scrap, weighed it, and then backed the cars up to the other
bins and had several tons of No. 1 scrap thrown on before making his
purchase. The bills were only made out for No. 2 scrap. Jewett also
claimed that Dringer managed to purchase 1,700 tons of old Ramapo
railroad car wheels below fair market price. E. S. Bowen, general su-
perintendent of the railroad, testified that Dringer made a bid of twenty-
two dollars per ton for the wheels, that he subsequently receded from
the offer, and that Colonel Bowman sold them to him for nineteen dollars per ton.

Jewett's allegations, sustained by numerous affidavits, convinced
New Jersey Chancellor Runyon to issue an injunction on 1 April 1876
forbidding Dringer from disposing of any of his old metals and placed
his yards in the hands of a receiver. Dringer was ordered to account
to the Erie Railway for all his dealings with the road. Colonel Bowman was forbidden to leave New Jersey, and he was also ordered to
answer the charge of conspiring with Dringer to defraud the company.[10]

The resolution of this case ultimately favored Dringer. The charges
were thrown out, and he was permitted to sue the railroad for lost revenue. The dealer and his customer exchanged several lawsuits over
the next decade. The mistrust surrounding Jewett and Dringer indicated how contentious dealings could be in old materials, with oral
agreements and lack of standards opening negotiations to charges of
fraud and dishonesty.

The Erie Railroad's suspicion of price fraud was not isolated. Other
customers were worried that scrap dealers falsified weights or qualities of goods. In a 1914 article in *Metal Industry*, W. H. Parry, super-
intendent of the National Meter Company in Brooklyn, warned
customers of scrap dealers "to be on the constant lookout, or some
pretty raw deals will be slipped over on them that will prove at least

why they prefer to be called 'scrap dealers.' If there is any one trick that these dealers do not know of I have yet to hear of it."[11]

The schemes Parry described included one involving a copper dealer who froze water inside copper casks he intended to sell in order to inflate their weight. Another story involved a talented 150-pound Saint Bernard that was so well trained that "if his boss is selling materials he just loves to be weighed in with the load. He also has a strange antipathy to weighing out, but let his boss be the buyer, he then insists on being weighed as part of the empty truck, that is, to convey the materials outside the premises, but never as part of the truck gear when the load leaves."[12]

Parry also alleged that dealers employed "the old dodge of weighing the driver plus a couple of well-weighted feed bags," manipulating the balancing weight of truck scales, and having truck drivers "place one of the truck wheels so that half of its width, at least, is not on the scales."[13]

Lest one suppose that Parry was unusually harsh in his description of scrap dealers, he had plenty of company making allegations. A 1914 article in *Technical World Magazine* reported that "the old metal business is probably the crookedest of any that is classed as legitimate" and one where collusion among merchants was common. A reporter writing to readers representing the U.S. War Department warned that scrap dealers colluded under a gentleman's agreement not to bid against each other in order to get copper worth thirteen cents a pound at one or two cents a pound and other material at proportionate rates. "Also, copper weighed on the junk man's scales weighs but six or seven ounces to the pound when purchased, though the same metal runs twenty to twenty-five ounces to the pound when sold, if the buyer is not wise to this old method of proceeding."[14]

Often these complaints were couched in rhetoric that conflated the scrap dealer's unethical business practices with his status as a mysterious, unscrupulous foreigner related to the shylock characters perpetuated in European and American folklore since European Jews were accused of usury in the Middle Ages. By the late nineteenth century, the crooked Fagin in Charles Dickens's *Oliver Twist* represented popular perceptions of the child-corrupting, thieving Jew. Official correspondence and trade journals were more reserved in their criticism, but their wariness was explicit. J. P. Alexander, writing in the *Electric Railway Journal*, warned scrap buyers that "as the 'scrap business' is carried on to a great extent by foreigners, and classes of collectors who are constantly going beyond the limit of the law, all sales where such people become bidders are best handled on a strictly cash basis, and with an exact agreement in writing as to the details of each sale."[15]

Such warnings were made against generic foreigners most of the time, but specific accusations against Jews indicated how much both Jewish identification with the scrap trade was complete during this period and how much anti-Semitism pervaded American business practices. Nelson Whitaker, writing to his attorney in 1895 regarding a firm he felt was attempting to defraud his Principio Foundry Company, emphasized that they must be cautious in handling the situation, as he was dealing with "a firm composed entirely of Jews."[16]

Popular depictions of peddlers and dealers emphasized their ethnic dimension, linking ethnicity to unscrupulous behavior. The Polish Jew Zerkow in Frank Norris's 1899 novel *McTeague* is an extreme example. He calculates the value of everything he sees and marries a woman because he mistakenly believes she possesses gold dishes. Obsessed, he is driven to delusion attempting to track down the elusive gold. The popular image of the junk dealer at the turn of the century was a hook-nosed, scheming Jewish immigrant bent on cheating honest Americans.[17]

Scrap dealing could be volatile and unpredictable; prices of materials fluctuated rapidly and competition was fierce. Many dealers resorted to desperate means to gain advantages over their competition. Such short-term efforts to make quick money undercut the long-term viability of the industry, but as one veteran broker put it many years later, "that generation was just trying to stay alive," and some individuals did whatever it took to stay in business.[18] Novelist Mordecai Richler later immortalized this mercenary attitude in his novel *The Apprenticeship of Duddy Kravitz*. Montreal-based scrap dealer Moishe Cohen tells the protagonist that all successful businessmen had "something locked in the closet" regarding their past business dealings and that the young man had to have the ability to "make himself hard" to the realities of business, be ruthless, and succeed. For Cohen, this meant squeezing every penny out of his customers and investing in cheap equipment that might get his employees killed. Such ruthlessness contributed to the negative depiction of scrap dealers, but given the cutthroat competition in the trade, a desperate dealer might feel it was necessary to survive. Survival was tenuous and depended upon both skill and good luck. Though many men made their fortunes in scrap at the time, many more were unable to harness the requisite skills, good fortune, and ability to overcome the unscrupulous elements in the industry and left the business.[19]

Even those who had succeeded in the scrap trade, such as Sam and Abe Levinson, grew weary of attacks on their character as junk dealers. Once the brothers found viable alternatives, they left the trade for more reputable occupations. In Sam and Abe's case, abandoning their father's scrap business led to successful careers in steel fabrication. Levinson Steel became one of Pittsburgh's most successful steel firms, and the Levinson family became respected as community leaders and philanthropists. Whether the family would have risen to such social prominence had they remained in scrap is unclear; that the brothers decisively left the trade despite their successes indicates they preferred the risks of entering the steel industry rather than remaining in scrap.[20]

Customers wishing to avoid the chicanery of unscrupulous scrap dealers had other options. They could attempt to salvage more of the scrap generated by their own operations rather than purchasing scrap on the open market. Between 1910 and 1920, engineers employed by large companies exhorted their employers to develop more rational, efficient means of reclaiming their scrap. R. P. Warner, vice president of Griggs, Cooper and Company, claimed in 1917 that waste was "extremely insidious in undermining and reducing the profits of a business," and the problem was serious enough that "the general character of a whole business may be sized up pretty accurately, and in a large measure be kept track of and regulated, by observing closely the waste from each department."[21] The impulse to reclaim wastes was rooted in a notion that all waste could be reused if engineers devised efficient means to claim and process it. Theodore Waters claimed "every possible substance we use and throw away comes back as new and different material—a wonderful cycle of transformation created by the scientist's skill." The "waste to wealth" school of thought assumed that all industrial and postconsumer material had economic value and that maximum economic efficiency depended upon proper reclamation procedures.[22]

The drive toward maximizing the use of materials conflicted with the growing tensions in the household to dispose of everything that was not new and clean, but it proved popular in industry. Foundries, mills, and railroads developed programs to reclaim and reuse their scrap. The Union Pacific Railroad Company estimated that it saved more than one thousand dollars a month in its Omaha, Nebraska, shops just by reclaiming its old bolts and nuts.[23]

While engineers stressed the virtues of conserving materials in the factory, efficiency measures were not adequate for steel firms desiring large volumes of scrap. Those firms wishing to purchase their scrap—a segment of the market that included mills and factories whose needs were beyond the most efficient reclamation of in-house scrap—could purchase old materials directly from railroads, other mills, or military auctions instead of relying on scrap dealers to act as skilled middlemen. This practice was known as direct dealing and required the buyer to take the responsibility to find suppliers that could provide adequate supplies of high-quality scrap at desirable prices. Direct dealing was easier said than done, as the scrap trade had evolved into a complex, informal network of several small dealers supplying larger brokers. Those larger brokers had the ability to fill industrial customers' orders with scrap purchased by dozens of smaller dealers, purchasing from different yards depending on available stocks and prices.

Most mills used scrap brokers because brokers depended upon a complex series of relationships that could generate supplies and prices that a novice could not hope to match. Many felt that using brokers was simply easier than trying to learn how to duplicate their services. Some, however, felt the risk of fraud was too great and opted to bypass the scrap trade. Bethlehem Steel engaged in direct dealing in the 1920s when it opened its own scrap yards and purchased abandoned ships and machinery directly from sources such as the U.S. Navy, railroads, and other mills, bypassing brokers. Scrap dealers were alarmed at Bethlehem's moves and feared that direct dealing threatened scrap dealers with the loss of their largest customers and suppliers.[24]

A Moral Menace

Concern over customer satisfaction, though significant, was not the only reason why scrap dealers needed to worry about legitimacy; a business perceived as illegitimate encountered conflicts with other segments of society. Worries about hygiene had allowed immigrants to establish businesses in the waste trades, but they also aroused worries about the new businesses. The physical hazards posed by scrap materials, along with suspicion that the immigrant businesses practiced fraud and invited violent and economic crimes led Progres-

sive reform groups and law enforcement officials to regulate or eradi-
cate scrap businesses.

Public health professionals had moved away from the miasma
theory of infection in the late nineteenth century in favor of the ex-
planation that microscopic organisms produced infectious diseases.
The new model of disease transmission had several implications for
dealing with epidemics, including preventive vaccinations and the removal of matter infected with communicable diseases. Municipal health
departments and voluntary organizations in both Europe and the United
States focused on shielding vulnerable people, especially children, from
infection. Nurses taught residents to avoid infectious material and wash
regularly. In the late nineteenth century, public health institutions such
as state boards of health, municipal sanitation departments, the U.S.
Army Medical Corps, and professional organizations such as the American Public Health Association concluded that individual responsibility and behavior for disease prevention was insufficient and that
systematic prevention by government bodies was necessary.[25]

Rags, especially those imported from Europe, were now seen as
carriers of infectious diseases and were often quarantined by the port
of New York in order to prevent epidemics. In addition to their suspected role in causing disease, the incendiary nature of rags caused
fires in many shops, endangering neighboring buildings, residents, and
firefighters. The *New York Times* reported concerns over fires, small-
pox, scarlet fever, and other health threats associated with rags between
1850 and 1930, with concern over diseases escalating after 1890 as
bacteria-infected cloth became perceived as a serious danger to public health.[26]

Large cities such as New York, Philadelphia, and Chicago gave increasing scrutiny to old, potentially infected matter, though disease
transmission was not the only concern municipal officials had with
scrap materials. Iron and steel scrap posed fewer immediate threats
to physical health, yet Progressives working to improve urban life
blamed the scrap metal trade for a variety of moral and physical ills.[27]
After World War I the Chicago Juvenile Protective Association, one of
many Progressive efforts to improve the welfare of urban children, issued a report entitled *Junk Dealing and Juvenile Delinquency* assessing the impact of the city's dealers on Chicago's children. The
association echoed many customers' complaints in charging that junk
dealers were "low in the scale of ethics and intelligence." In addition,

it concluded that junk dealing was dangerous, promoted theft and violent crime, and was "a moral menace to minors . . . fraught with grave dangers to health and physical well-being."[28]

The association singled out the junk trade as a danger to children because it felt the dealers encouraged theft, taught the lesson that one could put a price on any object, and engaged in a trade that was highly competitive, disorganized, and prone to violence. The association was primarily concerned with the relationship between children and junk peddlers using wagons, whom the study estimated had doubled in their number during World War I to about 1,800 canvassing Chicago. These peddlers collected scraps and sold them to the increasing number of retail junk shops and wholesale scrap yards proliferating in poorer urban neighborhoods, and it was in this environment where they came into contact with children.[29]

For decades prior to the association's report, Jane Addams had worked in Chicago to improve the lives of urban residents. Her opinion of the junk trade shaped the association's concerns. In 1899 the Hull House founder observed "there are certain boys in many city neighborhoods who form themselves into little gangs with leaders somewhat more intrepid than the rest." The gangs broke into vacant houses, where their "favorite performance" was to strip the houses of faucets and lead pipes, which they sold to the nearest junk dealer. Once they paid, the boys would buy beer, "which they drink in little freebooters' groups sitting in an alley. From beginning to end they have the excitement of knowing that they may be seen and caught by the 'coppers,' and at times they are quite breathless with suspense."[30]

Such concerns were neither unique to Chicago nor new to the waste trades in 1920. An 1873 case in New York City concerned a junk dealer who received stolen iron from boys. Iron manufacturer Martin Briggs testified against junk-shop owner Stephen Coleman, whom he accused of having iron stolen from Briggs's business, telling Coleman that he ought to know that when pig iron was brought to be sold by boys they must have stolen it in all human probability, and "we ought to be careful of buying of boys." According to Briggs, Coleman ignored him, claiming he "did not think it his business to look around and find out where it came from," and despite repeated requests from Briggs, Coleman remained in business and continued to buy iron from boys who presumably stole it.[31]

Jacob A. Riis observed the problem of junk dealers purchasing stolen goods from boys had not dissipated in New York City in 1894. The

Figure 3. Close housing adjoining junk, Milwaukee, Wisconsin. Scrap storage like this spurred zoning restrictions after World War I. *Photograph by Carl Mydans. Courtesy Library of Congress, Prints and Photographs Division.*

photographer saw soda-water wagons as easy marks for young thieves beginning their descent into lives of crime. "The bottle is worth so much cash at the junk-shop," so boys swipe them from merchants. "It is not a very great crime, but it is the stepping-stone to many greater. A horse-blanket or a copper-bottomed boiler may be the next thing. It is the first step that costs an effort, and that not a very great one, with the clamor of a hungry stomach to drown the warning voice within him that whispers of the policeman and the lock-up."[32]

Progressive reformers attacked the aesthetic nuisances associated with scrap yards. As most scrap men were immigrants living in urban areas, their yards were often located in residential neighborhoods. Many stored their goods in their front or back yards in plain view of neighbors and onlookers. The association report noted that scrap dealing out of residential neighborhoods posed a health risk to the area's children. "Boys hang about the shop while junk wagons are being unloaded or play in dirty piles of rubbish heaped up in the street."[33]

The threats to health, children, neighbors, and customers produced municipal efforts to control or eliminate the dangers posed by scrap dealing. Cities established zoning regulations to regulate where scrap yards and junk shops could be located, forbidding them in residential areas. In 1918, Philadelphia banned junk shops from residential and business districts and buildings used as dwellings, two-family dwellings, rooming houses, or tenements. The chief of the city's bureau of health's division of housing and sanitation explained that these ordinances were passed to place restrictions "upon those persons handling waste without any respect for sanitary law or for the health and comfort of their employees."[34]

Many cities licensed waste dealers and peddlers in an effort to get them to comply with local sanitary and trade regulations, and zoning laws prohibited their activities in some areas altogether. By 1920, municipal governments viewed the scrap industry as a threat to the public welfare, and many customers saw scrap dealers as unsavory elements. The industry had a low status and poor public image associated with physical and moral ills.[35]

Essential and Legitimate

Faced with the loss of their customers through direct dealing, and with the possible loss of their business location through municipal zoning, scrap dealers attempted to secure their legitimacy. Scrap firms mounted efforts between 1900 and 1930 to promote the interests of the industry. These efforts included establishing trade associations and journals, industrywide standards and ethical codes, and crafting rhetoric affirming scrap dealers' importance to the nation.

The *Waste Trade Journal* was one of the earliest national institutions established. While the iron trade publications had featured regular columns and articles on the scrap iron trade since the 1860s, the *Waste Trade Journal* was the first periodical devoted to the interests of scrap and rag firms. Founded by New York City journalist Charles H. Lipsett in 1904, it began as a weekly paper publishing the prices of waste materials in several markets and allowing firms to advertise their materials or solicit employees. Lipsett had expanded the journal by the early 1920s to include editorial content advocating the interests of scrap firms as well as supplying weekly prices for scrap metal, rubber, glass, and rags from several cities. As manufacturing journals such as *Iron Age* and *Iron Trade Review* questioned the ethics and importance of scrap firms, the *Waste Trade Journal* championed the trade.

Trade associations advocating the common interests of firms were commonplace by World War I. Steel producers formed the American

Iron and Steel Institute (AISI) in the 1870s, and dozens of industries organized similar associations in the late nineteenth and early twentieth centuries to establish industry standards and lobby on behalf of their members' common interests.[36] The scrap industry lagged behind the steel industry in forming trade associations. The first national trade association for scrap firms was the National Association of Waste Material Dealers (NAWMD), founded in 1913. The NAWMD represented brokers, dealers, and processors of waste materials including cloth rags, waste paper, scrap rubber, metals, and other materials, and its charter emphasized its quest in seeking the legitimacy of its members' activities:

> We feel and like to emphasize that the scrap-iron business is legitimate. We mean by this that there is a logical place for the scrap-iron dealer fitting between the producer and consumer. He should not be regarded as any unnecessary middleman or commission agent. He is as essential as the wholesale or retail merchant in any other line of trade. . . . It is therefore our duty and aim to tone up the general morale of the trade. We intend to encourage honest dealings and discourage dishonest and unscrupulous dealings.[37]

The NAWMD encouraged the regulation of fair and ethical trade among members, worked with regulating authorities and customers to prevent undue restrictions on trade and maintain goodwill, and promoted the trade as a vital, unique, and necessary industry.

The NAWMD claimed international membership by 1923, boasting members in Canada, England, Germany, Italy, Belgium, France, Switzerland, Holland, and Sweden. Most of these firms were involved in the paper, cloth, and rag trades, though scrap metal firms in Birmingham, England, and Italy were involved. Within the United States, most of the members were based in the New York metropolitan area (87 of 277 members listed in the 1928 association directory). New England boasted twenty-six members; Philadelphia twenty-five; and Chicago eighteen. Fewer than ten members were located west of Saint Louis, indicating that though the trade spanned the entire nation, it remained concentrated in the industrial Northeast. The largest trades advertised by the members in 1923 were in metals and paper mill supplies.[38]

By the time the NAWMD celebrated its fifteenth anniversary, many ferrous scrap firms wanted a second national trade association to represent their specific interests. In 1928, the Institute of Scrap Iron and Steel (ISIS) organized under the directorship of lawyer Benjamin Schwartz. ISIS differed from the NAWMD in that it specialized in the concerns of scrap iron dealers rather than the concerns of all secondary material dealers. Schwartz's immediate concern was the threat of direct dealing posed by Bethlehem Steel's new scrap yards, but the association's agenda was broader, including lobbying the federal government, creating a dialog with AISI, and improving relations among its own members. ISIS's founding pointed to the rapid growth of scrap iron as the nation's most widely traded scrap commodity by the end of the 1920s. Though its members shared concerns with the NAWMD's (indeed, many scrap firms belonged to both organizations), the scrap iron and steel firms concerns (notably the problem of direct dealing) led them to form a more specialized trade association.

The NAWMD and ISIS shared a common strategy of defining scrap dealing as a method of conservation. Conservationism grew in the early twentieth century as citizens worried about the fate of forests and undeveloped land in the wake of an industrialized and consumptive society; efforts to protect the nation's natural resources included the establishment of government agencies such as the U.S. Forest Service (which under Gifford Pinchot in 1905 assumed management of federal forests) and private groups such as the Sierra Club (founded in 1892 by John Muir). By the 1920s, conservationism became an emphasis of industry concerned with what historian Samuel P. Hays calls "the gospel of efficiency." The federal government and private industries worried about the finite availability of lumber, iron ore, and other resources that fed the economy.[39]

Under the Coolidge and Hoover administrations, citizens and businesses were urged to avoid what contemporary economist Stuart Chase labeled "the tragedy of waste." Chase argued that waste was a serious concern that hurt the nation's economic well-being. He claimed that waste was an engineering problem that could be solved by trained professionals working in industry and government.[40] Chase's argument resonated with industry and especially with Herbert Hoover's efforts as commerce secretary and president to minimize waste.

The new scrap trade associations reflected the growing concern over efficiency and conservation espoused by Chase. Theodore Hofeller, the Buffalo scrap rubber dealer who became the first president of NAWMD in 1913, stated "the waste material dealers are the true conservationists. They have reclaimed millions of dollars from the dump heaps throughout the country."[41] ISIS used the rhetoric of conservation to promote its members' virtues. Director Benjamin Schwartz consciously compared scrap dealers' work to the resource conservation advocated by President Hoover. Schwartz often quoted the president in his speeches and linked the scrap industry's well-being to the economic health of the nation, claiming that "many scrap iron dealers serve American industry not only by collecting scrap iron and steel, but by also gathering scrap rubber, scrap metal, rope, paper, and other waste materials," and that "the ramifications of a disorganized scrap industry go beyond the realm of the steel industry and that the effects on the national policy of conservation may be marked in a comparatively short time."[42]

A disorganized scrap industry, according to Schwartz, was the result of direct dealing and other practices that undermined brokers' abilities to acquire, process, and sell materials. A firm such as Bethlehem Steel that attempted to do its own trading would not only wind up getting inferior materials at inflated prices, but would also weaken the existing trade, causing expert brokers and dealers to go out of business and the American economy to lose their unique and important skills. In this view of the trade, concerns about fraud paled in comparison to the unmatched efficiency skilled scrap men provided in locating, processing, and supplying scrap materials to the nation's manufacturers.

This is not to say that the trade associations discounted the accusations of dealers as shysters. In addition to promoting scrap dealers as agents of conservation, the NAWMD and ISIS undertook several measures intended to improve their constituents' images. The associations prepared industry codes of ethics in consultation with customers and the Federal Trade Commission (FTC). These codes provided for specific guidelines on quality standards, independent arbitration of grievances, and penalties doled out to members who violated the codes.[43]

Trade associations worked with customers to establish uniform standards on what constituted grades of quality. While standards classifying materials into grades had existed since the mid-nineteenth century, disputes over what constituted preferable No. 1 grades of material over inferior No. 2 grades of material were frequent and contentious, often ending up in court. Written standards agreed to by the FTC, ISIS,

NAWMD, and manufacturing trade associations such as AISI codified grading standards in an effort to reduce disputes.[44] These standards allowed customers some measure of confidence that they were getting heavy melting steel and not lightly sorted bundles of No. 2 scrap steel (or, for that matter, casks filled with ice). The associations backed the new codes by mandating that dealers who breached the codes could face fines or expulsion from the trade associations. By 1938, when it published a Blue Book celebrating its twenty-fifth year of operations, the NAWMD threatened to publish the names of expelled members to expose them to the market as frauds. The punitive measures were designed to give customers confidence that association members were certain to be upstanding businessmen. Those dealers who were members could be said to be in compliance with the associations' codes, and those dealers who were not could be perceived as ineligible for entry or possibly expelled. The associations hoped to convince manufacturers that dealing with a member of NAWMD or ISIS would guarantee sound business practices. These steps allowed scrap dealers to gain some measures of trust with customers and regulating bodies.[45]

Scrap dealers used their institutions as a form of Americanizing the trade, bringing it out of the stereotype of scheming shylock immigrants and into the mainstream of American economic culture. Dealers distinguished themselves from unethical competitors by identifying themselves as modern, honest agents of conservation versus premodern, unscrupulous junk men. Schwartz made this comparison when attacking the practice of direct dealing done by Bethlehem Steel and other customers.

> The very foundation for the existence of the members of the Institute of Scrap Iron and Steel, which we have emphasized, is the service that the dealer renders to American industry. This service consists in the collection, the proper preparation and distribution of scrap on the open market, in accordance with contract requirements and in accordance with the needs of the steel industry. The development of the policy of buying only unprepared mixed scrap, undermines the very foundation of that service and *substitutes an exalted junk collector for the modern scrap iron dealer.* (Emphasis added.)[46]

The new scrap trade associations reflected the growing concern over efficiency and conservation espoused by Chase. Dealers stopped identifying themselves as *waste* material dealers, for "waste" was a pejorative word, one that distorted the scrap dealers' activities. A good scrap dealer, though he handled materials that had been discarded, *prevented* waste and maximized efficiency. "Scrap" and "secondary materials" were valuable commodities vital to the nation's productivity; "waste" was of negative worth and might pose dangers to public health. The beginnings of the scrap industry's professional associations began a long process of attempting to separate its activities from the notion of waste. The NAWMD changed its name in the 1950s, dropping the term "waste material" and opting to become the National Association of Secondary Material Industries (NASMI).[47]

The dichotomy Schwartz posited between the "modern scrap dealer" and the "exalted junk collector" not only attempted to assure customers but also implicitly addressed city residents' fears of the threats the junk trade posed to urban moral and physical health. According to this dichotomy of images, the junk collector in 1930 was corrupt, foreign, mysterious, and a threat to the public welfare. Conversely, the modern scrap dealer was a rational, efficient, honest, patriotic businessman concerned with the nation's welfare, especially as he saved valuable resources from being wasted. The junk collector polluted city streets and neighborhoods with unsightly piles of rags, old metal, and other debris infringing on the sensibilities (and health) of urban dwellers. The modern scrap firm did its business away from residents, in an organized, technologically modern industrial setting, where it performed its valuable work without creating nuisances or threats to public health.

The distinction—though exaggerated—held some truth in 1930. Much of the criticism of aesthetic and moral ills had to do with scrap businesses' proximity to residential areas. Most of the largest yards were located on the urban periphery or in industrial areas away from the residential neighborhoods where zoning ordinances had begun to ban their activities. This spatial pattern was due more in part to economic interest than compliance with zoning ordinances. The largest scrap customers were located in such areas, and access to warehouse space for large loads was available in industrial areas. In Pittsburgh, industrial and residential activities both left downtown to (largely) separate areas of the city, leaving the central business district near the point as a financial center. Scrap firms moved to the Strip District, close to the steel mills on the river. Scrap yards often located near railroad tracks, allowing shipment of large loads to and from the yards, and many yards located at ports to rivers, lakes, and oceans, allowing for affordable water transportation.

The smaller yards, shops, and peddlers that supplied the large brokers did not disappear, and their activities could not be eliminated from residential areas, especially since peddlers depended on contact with residents to collect their materials. As zoning ordinances proliferated, these businesses survived by exploiting the regulatory inequalities produced by the economic and racial segregation that grew in most American metropolitan areas between World War I and the 1950s. As the compact walking city of the nineteenth century expanded into the commuter metropolis of the mid-twentieth century, the larger urban area became divided so that the middle-class was now separate from the poor and African Americans became more trenchantly segregated in declining inner city neighborhoods where they had the least access to road repairs, policing, and other public services. Smaller yards and shops persisted in poorer inner-city neighborhoods, where they were close to postconsumer sources of scrap and might be overlooked by lax zoning enforcement. As the mass migration of African Americans to northern cities during World War II combined with mass suburbanization of whites after the war to expand ghettos, these establishments concentrated in African American neighborhoods. Gerald A. Gutenschwager observed in 1957 that scrap firms concentrated in Chicago's South and West Sides between World War I and 1956. Those areas had become among the poorest and most racially segregated in the nation since 1920. As with prostitution, gambling, and other vices cities sought to limit, public officials pushed scrap dealing into areas with little political power rather than eradicating the activity within the city limits. Despite steps to improve the industry's image, scrap trading remained an undesirable activity that shared moral and physical space with other undesirable activities. Though this pattern represented growing environmental inequalities between rich and poor and white and black, it allowed smaller businesses in the trade to persist despite attempts to eliminate them from urban life.[48]

Junk peddlers canvassed some neighborhoods as recently as the late 1940s, though they were in decline after 1930. These businesses, some consisting of little more than collections of scrap metal piled in a dwelling's front or back yard, were the source of many complaints from neighbors and police. Peddlers roaming residential streets with wagons of miscellaneous junk also drew complaints. The negative public images associated with scrap typically focused on the smaller businesses in the industry, as they were more conspicuous in everyday urban life.[49]

The larger scrap dealers could disassociate themselves from their smaller contemporaries, though they also did business with them, purchasing smaller collections from peddlers and inner-city yards in order to fill out large orders from their own customers. Many scrap dealers loathed their contemporaries' dishonesty, not simply because it gave the occupation a bad reputation but also because they themselves risked becoming the victims of fraud. Thus, arbitration of disputes by association officials was an important aspect of associational life, attempting to give dealers some measure of protection against unscrupulous partners. Such business dealings and conspicuous, negative profiles of peddlers and inner-city dealers caused problems for larger dealers. The dichotomy scrap dealers set up between their modern, upstanding image and one they characterized as antiquated was intended to distance themselves from negative images held by customers, regulators, and the public.

Aspects of the industry remained volatile, but despite occasional periods of depressed demand the industry continued to grow and diversify.[50] The threat of direct dealing was of primary concern to independent dealers. Schwartz claimed the skills required of the scrap dealer made him a unique and vital part of the nation's economy, one that could not be adequately supplanted as customers attempted to do by direct dealing. The "steel mills cannot gather and prepare scrap as economically as the scrap iron dealer. Those who have attempted to do so in the past and have kept track of their experiences, have abandoned the practice," he said. Though mills invested in the same technologies scrap dealers did, and had more capital to spend on such investments, scrap dealers argued that they had developed a specialized expertise that could be neither replaced nor ignored.[51]

The trade associations' claims that scrap dealers were modern, ethical agents of conservation with unique and vital skills became standard rhetoric in the industry's lobbying efforts. ISIS and NAWMD representatives frequently testified in congressional hearings, influencing legislation related to the industry.[52] Though friction between scrap dealers and their customers did not disappear, larger brokers such as the Luria Brothers enjoyed long-term relationships with the largest American industrialists, such as the Ford Motor Corporation and U.S. Steel, relationships built on trust, understanding, and a mutual recognition of dealers' value to their customers.[53]

Dealers' successes in rehabilitating their images, however, were not total. ISIS's campaign was unsuccessful in eliminating direct dealing; Bethlehem Steel opened several more scrap yards in Pennsylvania and Maryland between 1930 and 1950, including a giant facility at Sparrow's Point, Maryland. Yet despite engaging in direct dealing, Bethlehem simultaneously purchased scrap from independent dealers, and Lukens, U.S. Steel, Jones and Laughlin, and other large firms continued to purchase scrap from independent dealers through the 1930s and during and after World War II. Furthermore, Bethlehem Steel's success in operating its own scrap yards was unique in the steel industry, as mills found it very difficult to compete with experienced dealers for optimal prices and quality.[54]

The trade associations' efforts to advocate the virtues of scrap dealers did not prevent the establishment of new local ordinances designed to limit the industry's nuisances. Statutes such as Philadelphia's ordinance banning junk sales in residential buildings remained on the books. Dealers attempted to fight local licensing measures but were rebuffed by courts deciding that junk and scrap businesses posed too great a risk for theft and other crimes. A Kentucky judge, making such a ruling in 1930, directly addressed the scrap industry's claims to legitimacy, acknowledging that "it is, to be sure, a legitimate business, meeting a public demand, but it is the history of experience that it is sometimes conducted in a dubious fashion and becomes a place where thieves turn into cash their ill-gotten plunder." The ruling gave dealers the benefit of the doubt, stating that "it is, perhaps more often, an innocent receiver of contraband," yet as Briggs admonished Coleman in 1873, the court found that not knowing whether the source of the goods were ill gained damaged the dealer's reputation. "It is therefore a business having a potential danger to public peace, to public safety, and to public health. . . . The nature of the business supported the reasonable regulation contained in [Kentucky's junk] ordinance."[55]

The scrap industry, though it had fought for and achieved a measure of legitimacy by 1930, still faced problems relating to its association with materials mainstream American society regarded as dirty. The negative associations with waste and fraud remained, though now they were contested by a systematic response.

Vital Agents of Conservation

The battles scrap traders fought with customers, reformers, residents, and public officials reflected the social taboos they

broke as they established their businesses. As scrap firms became successful, they became conspicuous, and the low barriers to entry meant
thousands of competitors could enter the trade, some willing to engage
in deceit or objectionable activities in order to compete. As native-
born Americans attempted to make some sense of order of the turbu-
lent changes wrought by heavy industrialization, immigration, and
urbanization, they saw the largely immigrant businesses as threats to
the tenuous social order in industry and society.

Individuals engaged in the waste trades had their legitimacy questioned. Customers and suppliers criticized waste traders for not adhering to the conventions of accepted business practices and attempted
to reduce or eliminate business dealings with the new immigrant-
founded enterprises. Progressive reform groups and zoning offices in
many cities viewed junk dealing as both a threat to public health and
a cause of illegal activity and took measures to restrict the trade. Both
sets of objections saw the waste trades as threats to the American order of cleanliness as it related to germs and ethics.

The critics of scrap firms employed contemporary notions of waste
defined as chaos or dirt. Critics attacked the industry as an aesthetically unappealing activity that posed a threat to moral and physical
health as it introduced wastes and ill morals into residential neighborhoods. Scrap's association with waste materials gave it a low status and allowed immigrants the opportunity to enter the industry.
Distrust of foreigners conflated the industry's "dirty" status to include
disreputable business practices, producing a low moral quality in addition to the threats of disease and fire produced by storing and processing waste materials. The scrap industry's image of being unhealthy,
corrupt, and foreign threatened to hurt its ability to maintain customers and operate in the context of new municipal regulations.

In addressing those criticisms, the industry attempted to redefine
itself by identifying scrap dealers as fighters of waste—waste defined
not as unhygienic but as economic inefficiency. Industry members responded to these charges by employing rhetoric consistent with the
values of conservationism. Government and industry leaders espoused
conservation as a way to manage resources, and scrap dealers adopted
that concept, arguing that they provided a vital service to the nation's
economy as agents of conservation, that scrap dealers possessed unique
and important skills to provide that service, and that modern, honest
scrap dealers made up a legitimate enterprise that was an asset to both
industry and society. Contemporary notions of economic value—

notions that recast the commodity as *scrap* rather than *waste*—allowed the industry to claim its importance despite the aesthetic criticisms of its operations.

Scrap trade associations attempted to exert power over their members in an effort to minimize their customers' charges of unethical and illegitimate business practices. Through associations, dealers regulated themselves in order to present a convincing portrait of ethical business activity to customers and suppliers. Dealers' attempts at self-regulation allowed the industry to make claims of legitimacy, claims that responded to notions of national pride and sound business sense.

Though the scrap industry's legitimacy continued to be contested, scrap dealers were able to use them to establish relationships with customers and regulating bodies and continue their enterprises. Large firms complied with municipal zoning codes by situating in industrial areas; small ones evaded zoning regulations by situating in poor residential neighborhoods. Their experience represents an important dimension of how recycling functioned (and was perceived) in America's industrialized cities during a time when unprecedented levels of production and consumption made the activity a nuisance to some and a necessity to others.

By 1929, the scrap industry had developed a hierarchy of brokers, processors, dealers, peddlers, and scavengers working in loose networks to reintroduce scrap materials to industrial production. Scrap firms brought some semblance of order to the industry through the establishment of trade associations, journals, and standards of measurement. These institutions shaped the industry's response to criticism and were pivotal in making material reuse in 1945 quite different from reuse in 1929.

4 | All Us Cats Must Surely Do Our Bit

In September 1942, the Brooklyn Dodgers offered free admission to fans who brought ten pounds of scrap metal to Ebbets Field. In one day, sixty-five tons of mattress springs, bed frames, pots, pans, and assorted scrap metal were deposited outside the park by grinning fans lined up around the block to get into the game.[1] Almost a year later, Girl Scouts and other youth organizations across the United States collected scrap materials for the war effort.[2] An observer from 1928 would find these scrap drives puzzling. The people who handled scrap back then were either in private business, employed by the business, or scavengers trying to make enough to eat, not middle-class suburbanites and honor roll children. But times were different. Franklin Roosevelt had pleaded with all Americans to collect old rubber, copper, steel, and other materials and return them to the federal government, which would "salvage scrap to blast the Jap."

The federal government, through several agencies, including the Works Projects Administration, admonished Americans to save their scrap. Posters, articles, pamphlets, advertisements, radio shows, and public rallies attempted to mobilize the collection of tin cans, nylons, old cars, and rubber. These campaigns successfully galvanized mass action. Across the nation, thousands of voluntary drives led by businesses, fire departments, fraternal organizations, scout troops, and schools reclaimed household wastes, old industrial machinery, and

Figure 4. Campfire Girls collecting scrap metal, 1942. Juvenile scrap collecting for the war effort temporarily quashed concerns over children's exposure to waste materials. *Courtesy Nebraska State Historical Society.*

even Civil War artifacts. Scrap was on everyone's lips in the 1940s, including Fats Waller's. In his musical *Ain't Misbehavin'*, he encouraged "all us cats to surely do our bit" to save up "ev'ry little thing" for both the war effort and some extra money. The effort was a smashing success; at one point in late 1942, President Roosevelt felt the need to tell the nation that he was very grateful that so many citizens from all over the nation were collecting scrap (though he preferred that they not send it to the White House). Scrap collecting and sorting was a mainstream activity of interest to society and to the president of the United States.[3]

Just twenty years earlier, however, scrap collecting was seen as the scourge of the cities, something to be stamped out with zoning and diligent policing. Yet by 1943 it seemed most everyone in the United States was in agreement with ISIS President Benjamin Schwartz's claims that scrap collecting was a vital, patriotic act that all Americans should do. What had happened to America? Had the scrap industry finally claimed its place in the national culture?

In the 1920s, even the most optimistic scrap dealer would have seen such a shift in attitudes as remarkable, though firms trading in scrap were thriving. By 1929, many scrap firms had grown beyond the ranks of collectors, peddlers, and shops to become processing centers and brokers handling large orders from industrial customers. Two national waste material trade associations had been incorporated, with one, ISIS, devoted exclusively to scrap iron and steel concerns. The rise of mass consumer products and industrial production appeared to ensure large supplies of and markets for scrap materials.

This World of Toil and Trouble

An observer from 1933, however, would have found the scrap drives of World War II a cruel joke. The Great Depression had eliminated most markets for scrap materials, throwing hundreds of scrap dealers out of business, their markets evaporating by heavy industry's lack of demand. In the plains states, an agricultural depression reduced farm machinery producers' demand for metal in the late 1920s, in turn depressing the market for midwestern scrap yards. In 1929, the stock market crashed, sending the nation's economy into depression and crippling industrial output. Steel mills, automobile manufacturers, construction firms, and most industries were forced to curtail production and purchases of materials. Piles of scrap metal lay unwanted for years.

The scrap industry operated under a grim economic context in the 1930s. The Great Depression kept manufacturers' demand for raw materials low, lower than it had been during the depression that followed World War I. Domestic prices of scrap commodities, especially iron, were consistently low. Sales volumes declined each year after 1929. Many small scrap firms failed or moved into other industries. The census estimated that net sales in dollars by all waste material dealers declined by 48 percent between 1929 and 1933, from $352,280,000 to $282,602,000. Specialists in iron and steel experienced a decline of 74 percent, from $258,794,000 to $65,801,00 over the same period.[4] Hundreds of small firms were wiped out as demand for their materials plummeted. Large firms fared no better. The Depression decimated scrap brokers' domestic sales. In 1929, the Census of Business enumerated thirty-nine brokers with $47,697,000 in annual sales. Four years later, the census enumerated eleven brokers had annual sales totaling $4,373,000. The gains made during World War I and the 1920s disappeared in the wake of a worldwide economic downturn.[5]

Economic disaster reshaped the industry. The businesses that survived the Depression saw changes in their operating structures, changes involving more complex management structures and sometimes changes in the firm's direction. These changes often occurred when founders' children became involved in the family business. Firm founders were often immigrants possessing no formal education. They performed or oversaw all aspects of purchasing, sorting, processing, and selling scrap. Second-generation scrap men were more likely than the founders to attend college, due in part to the strong emphasis Jewish immigrants placed on their children's education.[6] The younger scrap men received training as lawyers, accountants or engineers, bringing specialized skills back to the firm. Schiavone-Bonomo, a large New York–based scrap firm founded by Italian immigrants at the turn of the century, expanded its management by hiring outside the family for specialists. The firm's most notable hire was Herman D. Moskowitz, who took over management of brokering purchases from smaller firms. Moskowitz, who entered the industry as a sorter at age thirteen, would be senior management at Schiavone-Bonomo and ISIS for decades.[7]

Smaller firms also saw changes wrought by younger generations. Iowa scrap dealer Joseph Alter's son-in-law Bernard Goldman was a lawyer by training; when Goldman succeeded Alter as president of the company, he diversified its holdings to include transportation holdings (and, much later, riverboat gambling).[8] The children of scrap men transformed family businesses in size, scope, and orientation, some building new, and perhaps more respectable, businesses on the initial achievements of their scrap-handling fathers.

Many of the changes wrought by founders' descendants concerned investments in processing and transportation technology. Technologies used in the 1930s were similar to those used by dealers in the 1910s, though their size, capacity, and cost grew. The first gas cutting-torch came into use around 1910; it remained a common tool for processing heavy grades of scrap iron.[9] Lifting magnets and skull crackers were common features three decades after their introduction to the scrap yard. By 1930, the baler—originally adapted from hay balers—was a widely used way of turning bundles of loose scrap into easily manageable bales. Discarded consumer appliances and automobiles became the source of vast accumulations of light sheet scrap by the 1930s, and balers allowed processors to handle light sheet scrap in an orderly manner. Rag and waste paper dealers also adopted balers to bundle their materials.[10]

Scrap yards adjacent to rail tracks had advantages in shipping loads to customers. Smaller dealers and some peddlers purchased automotive trucks to transport lighter loads. Aside from a few minor developments, the technology used to process and transport scrap materials in the 1930s reflected continuities with earlier uses rather than innovation. Given the economic climate of the decade, conservative investments were understandable.

Though the technology used in the scrap industry did not evolve substantially during the 1930s, the demographic characteristics of individuals working in the industry changed. The number of foreign-born individuals in the industry declined. The demographic changes in the nation, produced in large part by immigration restrictions during and after World War I, consolidated another trend in the industry that had been developing since 1920. Firm owners and management were often the American-born children of Jewish immigrants, even outside the major cities. Mack Cottler moved to Texas as a young man in the middle of the Depression. Years later, he recalled that though there were very few Jews in the small towns he visited, almost all the Jews were in the scrap business. Cottler received many invitations to dinner, frequently from families who had daughters eligible to marry one of the rare Jewish boys who came into town.[11]

Siting of the trade and demographic changes in American cities led to a growing reliance on African Americans as labor. Numbers of new immigrants dwindled in the 1920s and 1930s. Natives and second-generation immigrants pursued more pleasant occupations than sorting scrap. As zoning practices pushed scrap yards into African American residential neighborhoods, African Americans became the sorters, cutters, and haulers.[12]

Though many African Americans worked for scrap firms, they did not own or manage large processing yards or brokerages, likely owing to racism among customers and other dealers. Jewish scrap dealer Leonard Tanenbaum recalled several instances of buyers and sellers abusing his African American employees if they even drove his truck to deliver materials to customers. African Americans faced barriers to upward mobility in the industry beyond those that Jewish entrepreneurs were able to overcome. Despite these barriers, African Americans in both the North and South were active participants in the economy of reuse.[13]

The Great Depression underscored the relationship of salvage to the needs of impoverished people. Though the trade in scrap materials declined sharply after 1929, and the number of scrap firms enumerated in the Census of Business also declined, the number of individuals who attempted to scavenge materials for subsistence multiplied in American cities. In Harlem, a colony of fifty African American men formed a makeshift community of scavengers at 134th Street and Park Avenue. Workers for the Federal Writers' Project observed in 1938 that these men, who had been carpenters, painters, brick masons, auto mechanics, upholsterers, plumbers, "and even an artist or two" before the downturn in the economy, lived in "crudely fashioned cabins that were located in the holds of some old abandoned barges that lay half in, half out of the water." They canvassed the city in the middle of the night, "collecting old automobile parts, pasteboard, paper, rags, rubber, magazines, brass, iron, steel, old clothes, or anything they can find that is saleable as junk." The scavengers delivered their findings to the yard across the street from the colony, American Junk Dealers, Inc., who paid them for their efforts.[14]

Much like the immigrant women and children canvassing New York City's docks a century before them, these men used junk collecting as a means of survival. Just as their predecessors had throughout the nineteenth century, many of these scavengers ran afoul of the law in their pursuit of junk. The Federal Writers' Project interviewed a German-born itinerant nicknamed "the Baron," who collected goods for a junkman located on 48th Street in New York City. The Baron disclosed he was once hired by a contractor to remove copper from a building's roof. According to the Baron, when he climbed up to the roof and removed the copper, the contractor promptly stole it from him, leaving the Baron to fall two stories and injure his leg. The Baron was one of many men scavenging for survival during the lean years of the Depression, straddling the margins of legal and illegal behavior and providing examples for large scrap firms' claims of a contrast between themselves and the less scrupulous junk collectors.[15]

By 1935, American material reuse patterns had begun to resemble parts of the patterns of the last century. While reuse in the American home was not as pronounced as it had been in the 1830s, women were motivated toward thrift and extending the lives of food, clothing, and material goods as long as possible in a lean economy. Most of the people collecting old materials did so for subsistence, gamely working for yards that could pay them enough for their next meal. New firms that made a living off of salvage were few and far between.

Grave Perils

The outlook for the purchased scrap market began to improve after 1935. The New Deal aided sectors of the economy, such as construction, that had ground to a halt, yet the construction industry's demand for steel in 1935 was a fraction of its demand in 1929. Across two oceans, however, two nations began sustained military investments that exceeded their domestic supplies. Japan, long poor in natural resources, and Germany quickly became major export centers for American scrap iron dealers after 1932. The Japanese iron and steel industry began to climb out of the Depression. Prices of scrap in Japan rose as the industry's demand for scrap rose and exchange rates for the yen fell. Military demand for iron and steel throughout the 1930s and a relative lack of scrap supplies on hand made Japan an attractive market for international scrap exporters.[16] German and British demand for iron and steel led to increased exports to those markets in the mid-1930s.

Scrap firms able to engage in the export trade found new, growing markets. Luria Brothers, unlike many firms in the 1930s, actually expanded. Max Luria established the offshoot firm Luria Steel and Trading Corporation in 1935 in order to concentrate on exporting finished steel and scrap. He died in 1939 and his sons took over the branch, eventually splitting Luria Steel and Trading from the larger company. Luria Brothers in turn began competing with Luria Steel and Trading Corporation in the export market.[17]

Exports provided dealers with desperately needed business, but they came at a cost to scrap dealers' already contested image. As Germany and Japan began expanding through Europe and Asia at the end of the decade, brokers risked undoing their image of being vital to American interests. The irony of first—and second-generation Jewish immigrants aiding the Nazi war effort did not diminish caricatures of scrap men as scheming foreigners. Critics blasted scrap exports to Japan as it became clear that Japan was fortifying its military. When Japanese military aggression in Asia increased in the late 1930s, pressure to limit Japan's access to American resources grew. President Roosevelt ordered an embargo of exports of scrap iron and steel except to Great Britain and countries in the Western Hemisphere effective 16 October 1940.[18]

Paul V. McNutt, federal security administrator and former commissioner of the Philippines, denounced scrap trading with Japan on the eve of Roosevelt's ban. In a speech at Boston's New England Town Hall, McNutt attacked American scrap dealers for selling Japan 8.5 million tons of scrap iron, allowing Japan to build almost 500,000 tons of warships as "an amiable, freedom-loving nation is, without realizing it, exposing itself to grave perils for the sake of the junk industry." He warned of the danger of "powerful men-of-war steaming into Manila Harbor, into Guam, and the Hawaiian Islands and—if we dare to think of it—into San Diego and through the Panama Canal. Huge, sleek men-of-war, flying a foreign flag, and made from American junk." McNutt argued that the United States "must stop selling to Japan $30,000,000 worth of scrap metal every year to build warships which threaten our safety."[19]

Scrap dealers responded to McNutt's criticism by stating that they did what the government of the United States allowed them to do, and they did nothing that other industries did not do. Indeed, steel manufacturers and petroleum companies also increased exports to Japan in the late 1930s and came under similar attack on the eve of the United States' entry into the war. Once the embargo was enacted, dealers complied. One year after the embargo, American observers called it a success, perceiving a "crippling" effect on the Japanese steel industry. Such talk was muted after the bombing of Pearl Harbor.[20]

The most corrosive charges came after the attack on Pearl Harbor, when newspapers claimed America's old bridges and railroads had been used to fashion the ships, fighters, and bombs that killed American servicemen. Facing passionate attacks, scrap dealers attempted to minimize the effect of exports to Japan. A primer on the industry published in 1942 claimed not only that the amounts of scrap sold to Japan over the previous decade were insignificant, but also that American scrap may have actually injured Japan's war effort. Had Japan not received American scrap shipments, the primer claimed, Japan would have certainly built blast furnaces designed to forge steel from what ISIS claimed was its ample supply of pig iron in Manchuko. Because, however, Japan had built open-hearth furnaces that used ferrous scrap— and United States sources of scrap were now gone—the time that American scrap dealers had sold to Japan had left Japan dependent on a source they could no longer rely upon, and that nation was "almost impoverished in that material" and "has been greatly embarrassed."[21]

Attempting to shift the onus of exactly who had been embarrassed by the dealings with Japan was one of many ways in which dealers attempted to rehabilitate their image during the war. Restating deal-ers' function as agents of conservation promoting the national inter-est took on new meaning as the nation mobilized to fight. The scrap industry's relationship with the federal government was closer and more extensive than it had ever been. Scrap firms coordinated their activities with federal agencies, and the federal government regulated the flow and price of materials in an effort to manage a wartime economy. The results produced a relationship that did not lack for con-flicts as the war progressed.

War Production

World War II dramatically reshaped the American economic landscape. Changes in the national and global economies again changed Americans' perceptions of material reuse. The state be-came involved in the market for scrap materials during the war, regu-lating sales and distribution in an effort to maximize reclamation efforts at the lowest costs. Manufacturing industries, already the agents of demand that spurred the creation of hundreds of scrap firms across the nation, began participating in the collection and sorting of mate-rials in a more active manner during the war, blurring the lines be-tween scrap dealers and their customers.

State intervention had the most dramatic effect on the business of buying and selling scrap during the war. Aside from the obvious curtailing the war had on export markets, mobilization of the nation's military industrial production led to extensive government interven-tion in industrial affairs. Scrap collecting, which had for most of Ameri-can history been the concern of private enterprise (and private enterprise that was seen as rather unsavory by respectable society), became a subject of intense interest for the federal government. Be-tween 1940 and 1945, federal agencies, particularly the War Produc-tion Board (WPB) and its Office of Price Administration (OPA) attempted to reshape the cycles of material collection with implica-tions for both American society and the scrap industry.

The immediate effects on the scrap industry were mixed; though the fruits of their labors grew in perceived value, their methods and markets were questioned and altered. The United States mobilized the nation's industrial resources to prepare for the possibility of entering the war, including petroleum and primary and secondary materials. In May of 1940, President Roosevelt established a National Defense

Advisory Committee to advise him on problems of industrial production, materials supplies, labor, agriculture, consumer protection, and price stabilization. Roosevelt gave Leon Henderson, the commissioner for price stabilization, authority to regulate the prices of the nation's raw materials in addition to Henderson's capacity to advise the president. On 11 April 1941, Roosevelt issued an executive order creating the Office of Price Administration and Civilian Supply, naming Henderson administrator of the new agency. In August, the civilian supply responsibilities of the administration were transferred to the Office of Price Management (OPM), and Henderson's agency became the Office of Price Administration (OPA).[22]

Initially, Henderson elected not to regulate prices of scrap iron and steel. Low domestic demand during the depression had produced abundant supplies of dormant stock readily available for military production. As military production over the summer and fall of 1940 increased consumption, however, prices of iron and steel rose. The price of No. 1 heavy melting scrap steel increased from an average of $18.46 per ton in August to $22.63 per ton in December. Producers of pig iron raised their prices as much as $2.00 per ton in October.[23]

Henderson was concerned that rising prices would curtail or slow defense production. He invited members of the steel producing and scrap iron and steel trade associations to assess how the OPA could best distribute steel to national defense without making unreasonable demands on producers and dealers. AISI sent five men representing the steel producers, and ISIS sent five men representing scrap iron and steel dealers to a meeting on 8 October 1940. The attendees agreed to set voluntary price limits, but within months these limits proved unsustainable and the voluntary cooperation had disintegrated. On 6 January 1941, the Price Stabilization Division of OPA convened another meeting with scrap dealers to restate the government's desire for low, stable prices. Leon Henderson threatened dealers that the next limits would not be voluntary. "Unless voluntary action is successful, drastic steps looking toward control will be recommended."[24]

Faced with the possibility of government control, scrap dealers once again attempted to set voluntary price ceilings on themselves. Brokers promised to keep prices stable in exchange for guarantees that the OPA would make sure that suppliers, including mills and railroads, would also adhere to maximum prices in order for dealers to remain in business.[25]

The Price Stabilization Division decided to study price relationships for different grades of scrap in order to understand why the voluntary limits had disintegrated in December of 1940. Using prices published in *Iron Age*, the division had assumed that if it fixed the price of No. 1 scrap iron at a set figure, then other grades would stay at ratios similar to their value relative to No. 1 scrap in a free market. That assumption proved wrong, and values of lower grades of scrap rose quickly.

Dealers and consumers, aware that voluntary controls were breaking down and that rumors that the division was moving to set prices, requested that any price schedule OPA implemented have input by dealers. Henderson agreed. After a 9 March 1941 meeting, the division took suggestions from dealers, considered its own research of prices, as well as the advice of ISIS and steel producers, and drew up Price Schedule No. 4 on 3 April.[26]

Price Schedule No. 4 attempted to cover all the major grades of scrap and geographical differentials. This meant a detailed appendix arrangement with specific prices for different grades based on a rigid delivery point price. The division unintentionally prevented sales from points at a great distance and curtailed sales at shorter distances because the set prices were inflexible. Revisions to the schedule followed in an effort to make it more effective.[27]

Federal officials began observing a schism between scrap dealers. With all of the small competitors, scrap was never the most unified of industries, even after the establishment of trade associations to mediate disputes and lobby. During the war, the divide between large brokers and small dealers was pronounced enough that federal officials wondered if they had erred in considering ISIS the legitimate representative of the scrap industry. OPA official John Hart noted that though the association represented 95 percent of the scrap industry in terms of enrollment, he suspected that the trade organization did not equally represent all of its constituents. Hart alleged that the actual power of ISIS was centered in the hands of a few large brokers such as the Lurias and Schiavone-Bonomo, who directed its policy and made the recommendations to OPA regarding federal policy concerning scrap. Consequently, Hart felt small dealers and collectors' interests were neglected. The OPA subsequently tried to bypass the large brokers by surveying opinions from small dealers irrespective of the opinions of ISIS.[28]

Though scrap iron and steel were the most closely scrutinized scrap materials the OPA regulated, the agency also intervened into the markets of several other commodities. In February 1941, the agency warned scrap aluminum dealers that it would not tolerate hoarding; it threatened intervention into the scrap aluminum market if dealers did not lower their prices. Over the next month, voluntary limits did not satisfy the agency and it established formal price ceilings for secondary aluminum on 24 March 1941. The issue of Price Schedule 2 reduced by 20 percent the going market prices for scrap and secondary aluminum, prices that had exceeded those for virgin aluminum.[29]

The OPA concluded that voluntary controls by copper producers were largely successful but that it was more difficult to get the much smaller and less organized scrap copper and brass dealers to control their prices. The agency issued price schedules for brass mill scrap on 22 July 1941 and copper scrap on 19 September 1941.[30] OPA policy on zinc trading was similar; large zinc producers were able to establish voluntary controls to the agency's satisfaction, but the smaller scrap dealers had their prices set by the agency with a price schedule issued 31 March 1941.[31] Likewise, primary lead producers maintained low prices but scrap lead dealers could not, and formal price ceilings were established 13 January 1942.[32] By early 1942, all of the nation's scrap metal markets were subject to price controls instituted and enforced by the federal government.

Dealers found compliance with price schedules difficult not only because the regulations put a ceiling on their prices but also because the OPA did not adequately account for the costs dealers incurred, including rising wages and transportation costs.[33] Freight charges, especially over long distances, could prove too great to sustain scrap trading. The OPA initially set a system of zone-delivered prices with thirteen regions, presumably set on preexisting patterns of trade. The wartime economy altered domestic trade patterns, with new demands from the shipyards and mills on the West Coast for scrap iron and steel, as well as demands from the steel-producing centers in Chicago, Pittsburgh, and Birmingham for scrap outside of their normal hinterlands. Furthermore, freight charges assessed from 1929 statistics ignored the severe decline in traffic wrought by the Depression; domestic trade patterns in 1941 featured much lighter interstate traffic than had existed in 1929. Freight charges, coupled with price ceilings, discouraged scrap dealers from dealing with buyers outside of the markets they had maintained during the Depression.[34]

Under protest from ISIS, the OPA quickly abandoned its initial price ceilings. In June of 1941 the agency devised a new model where dealers' price ceilings depended on the lowest available shipping rates between them and their buyers. That September, the OPA granted dealers in remote, and typically ignored, regions such as Florida and the Mountain States maximum prices of five dollars a ton greater than scrap in the rest of the country to encourage circulation of scrap from the entire nation.[35] Paul C. Cabot, deputy chief of conservation in the Office of Price Management Purchases Division, met with eighty representatives of dealers in nonferrous metals, iron and steel scrap, wool rags, cotton rags, and scrap rubber in an effort to coordinate their assistance. Cabot appealed to dealers' patriotism, claiming that "a 20 percent increase in the annual amount of [copper, zinc, steel, aluminum, lead, and rubber] would mean an added volume of 4,551,800 tons contributed towards our war effort. It is our purpose to ask your advice and cooperation to the end of aiding us in bringing about such a result."[36]

The OPM, like the OPA before it, attempted to find a middle ground between the needs of the war effort and the needs of the dealers. The *Times* reported that the office asked waste material dealers to nominate representatives for an industry advisory committee to advocate dealers' interests. The committee worked with OPM to devise plans to increase the flow of scrap materials from all supplies, including the nation's homes.[37]

Save That Scrap!

Concurrent to its efforts at coordinating industrial production for the war effort, the federal government also mobilized local institutions such as schools, churches, and voluntary organizations to collect old iron, copper, rubber, cloth, and other materials from homes, dumps, farms, and any sources the citizenry saw fit to scour.

Scrap drives were part of a public campaign to extend the resources of the nation. During the war, domestic automobile production essentially ceased. Domestic tire manufacture stopped as manufacturers saved rubber for military applications. Housing production, slowly on the rise in the late 1930s, came to a standstill during the war as wood, steel, and other raw materials were redirected to military campaigns. The U.S. Mint replaced copper pennies with steel pennies to preserve supplies of the comparatively scarce metal.

Extensive marketing campaigns for scrap drives involved posters, print ads, radio announcements, and popular songs. Propaganda developed by leading advertising agencies and distributed by the federal government accentuated the idea that saving and donating scrap materials was a patriotic act. Racial caricatures of Germans and especially the Japanese in posters underlined the idea that material reuse was in the best interest of the nation, a sentiment resembling what scrap trade associations were arguing for the preservation of the private scrap trade. "Salvage Scrap to Blast the Jap" became a common refrain in propaganda, with graphic illustrations of how collected scrap metal could be used to fashion bombs that could smite the slit-eyed snakes (as the Japanese were portrayed in posters) threatening the United States.

There is some question as to whether the scrap drives had any more than a symbolic effect upon the collection and use of materials; even if the amount collected was insignificant, the cultural perception of collecting scraps was an important new development. Waste was no longer something to remove from view, and wasteful behavior was a problem that needed to be eradicated for a public good. Any materials, no matter how dirty, had to be reclaimed to wage war. In this context, Americans saw little problem encouraging everyone, even children, to collect germ-infested or rusty discards.

The new public ethic of reuse left scrap dealers bemused, and the war was a strange new time for the professional salvagers. Dealers participated, as did steel and other manufacturing industries, in consumer scrap drives to aid the war effort. Publicly, scrap dealers embraced scrap drives as extensions of the virtuous work they had been doing all along. ISIS released a pamphlet that declared that as the scrap dealer was unable to induce or accelerate obsolescence in factories, homes, and farms, he welcomed "the co-operation of government bureaus, salvage committees, newspaper publishers, householders, farmers, school children, and others who can bring to the surface material which can be spared and will serve a more useful purpose when made available by the scrap industry for remelting into iron and steel for the war effort."[38]

Privately, however, scrap dealers expressed reservations about the federal government's sudden interest in reclamation. Prior to the embargo of scrap sales to Japan and Germany, federal attention to the trade was minimal. As scrap drives became an important part of the war effort, federal regulation of prices and practices became a source of great controversy in the industry. Dealers complained that the OPA and OPM had set prices too low for them to stay in business. Furthermore, consumer scrap drives provided dealers with grades of scrap of insufficient weight and quality to be of much value and made scrap markets favorable to manufacturing industries at dealers' expense. Pittsburgh's dealers complained that scrap drives reduced the volume of scrap handled by dealers in the city by as much as 40 percent between July and August of 1942. Dealers did not want to trade the light scrap generated in collection drives, and shortages from railroads and mills gave them little heavy scrap to trade. That scrap which was available was subject to price ceilings imposed by OPA, limiting dealers' income.[39] The question of who received the materials collected from the public also rankled dealers. When steel mills participated in scrap drives, they negated scrap dealers as conduits between public and industrial discards and delivery to industrial manufacturers, making dealers redundant.[40]

The new federal attention to the nation's scrap markets meant that scrap dealers attempting to work within the framework of wartime regulations risked confusion and punishment. Lack of interagency coordination produced regulatory difficulties for dealers. The *Wall Street Journal* reported that officials of the Office of Production Management and the Office of Price Administration disagreed on scrap prices. "OPA officials contend that the present ceiling is not so low as to discourage dealers from seeking and collecting the metal. . . . Generally, OPM officials believe that higher prices would encourage collections of junk."[41]

Scrap dealers' primary objection to OPA regulations was that the agency set prices without considering dealers' costs, particularly shipping costs. The OPA took their objections into account. On 17 March 1942, the administration announced that dealers could add a freight charge of 6 percent to their prices to account for rising freight rates.[42] Despite such modifications to regulations, many dealers felt that the OPA had regarded scrap dealers as crooks.[43]

The OPA had the power to enforce violations of its price ceilings through the nation's federal courts. The office launched over 270 civil cases during the war against violators of its price schedules. In October 1941, Leon Henderson charged that "certain members of the trade" had "deliberately inspired" rumors that ceiling prices on iron and steel scrap would be raised in the hope of slowing down scrap collection and sales for their own eventual profit. Henderson threatened that "those dealers who are hoarding scrap in anticipation of higher prices may find themselves in an uncomfortable situation if their actions continue to hinder the progress of the defense effort."[44] In an effort to uncover illegal hoarding, the OPA began taking photographs of scrap yards in New Jersey, New York, Massachusetts, and Maine and then presented the pictures to the House Small Business Committee in January 1942.[45] Scrap dealers and customers taken to court included the scrap firms Glosser and Sons of Johnstown, Pennsylvania; Stalman Brothers of Williamsport, Pennsylvania; and the Hodes Coal and Junk Company of Lockhaven, Pennsylvania; steel firms included Jones and Laughlin and Allegheny-Ludlum.[46]

The OPA alleged that Allegheny-Ludlum purchased scrap at $21.00 per gross ton at a time when the legal ceiling was $14.50. Jones and Laughlin purchased scrap at $20.00 per ton when the ceiling was $15.00. In addition to ignoring price ceilings, scrap dealers allegedly "top-dressed" scrap, placing a thin covering of high-grade scrap over a load of inferior grade, a practice forbidden by Iron and Steel Scrap Schedule No. 4. The days of the scale-jumping Saint Bernard and frozen copper casks, it seems, had not ended.[47]

In the midst of this flurry of accusations and prosecutions, scrap firms and their customers charged that the OPA's actions were unfair and curtailed business. The Pittsburgh Steel Company claimed (after the OPA had obtained an injunction against it) that the agency's actions were extraregulatory and failed to comprehend the natural pricing of scrap.[48] Inland Steel claimed that when the OPA revised Price Schedule No. 4 buyers and sellers could have many interpretations of the definitions of materials covered by specific price ceilings.[49] By 1944, the OPA responded to shortages of scrap by removing several price controls, making the activity more profitable to spur scrap collection.

Federal intervention into the scrap market did not end with the imposition of price schedules and regulations, but extended to providing scrap firms with a federally owned competitor. The government attempted to speed the recovery of scrap into the war effort by forming a corporation to purchase scrap materials. War Materials, Inc., a subsidiary of the Resolution Finance Corporation (RFC, a federal corporation that built plants and infrastructure for the war effort) opened its first branch office in Pittsburgh in August 1942.[50] The RFC formed the company to purchase and sell to steel mills an estimated minimum of five million tons of iron and steel scrap. This scrap came through acquisition of obsolete, abandoned, and demolished structures as well as other sources. War Materials was assigned to move scrap to markets in cases where the total cost of processing and transportation exceeded OPA ceilings. Its activities provided dealers with a competitor that was not subject to the regulations they faced.[51]

The peak of federal intervention into the scrap industry was 1942. Supply conditions were considered critical between 1940 and 1943, and both regulation and the fervor of consumer scrap drives attempted to maximize reclamation. The OPA modified prices throughout the war to maximize supply, waiving restrictions on the sale and transportation of many materials at the end of 1943.[52] As the Allies made advances in both the European and Pacific theaters, they reclaimed large tonnages of battlefield scrap, easing the supply crisis at home.[53]

Supplies of battlefield scrap materials deemed unsuitable for reuse in the war effort grew to the point that the armed forces established segregation yards at ports to handle the material.[54] At the segregation yards, Army Service Forces checked the material to see if it was repairable property and if so sent it to installations for reuse. Forces checked to see that no explosives were present in any materials. Once safety and reuse considerations were exhausted, remaining scrap was sold to dealers, who then resold it to domestic customers.[55]

Benjamin Schwartz resigned as head of ISIS in 1938 and joined the war effort in the Bureau of Economic Warfare in June 1942. By the end of 1943, he was chief of the scrap metals section of the Foreign Economic Administration (FEA). Schwartz continued to work with domestic scrap dealers coordinating import and resale of battlefield and other international scrap. He reported that American scrap imports in 1943 came from thirty countries, reversing a trend in which scrap had moved out of the United States since the early 1930s. The combination of demand for the war effort and availability of accessible battlefield scrap abroad made the United States a net importer of scrap iron and steel.[56]

Scrap collection during the war, then, was a combination of renewed efforts from scrap dealers and manufacturers and newfound passion by government agencies and the general public. Despite price controls and new regulations, the heightened interest in scrap collection led to much activity in the markets by established dealers and newcomers alike.

Wartime Opportunities

The scrap industry operating during World War II continued the demographic trends of the 1930s. The industry was more specialized and less foreign-born than was the scrap industry operating during World War I. The demographic makeup of the industry, and of the nation, was much different in 1940 than it had been in 1920. Immigration restrictions instituted between the start of World War I and the mid-1920s reduced the number of foreign-born individuals entering the country, eliminating what had been a source of entry-level entrepreneurs in the industry between 1850 and 1920.

Trends in generational upward mobility affected the composition of the workforce. Many children of foreign-born parents received education and had access to networks of opportunity that allowed them to pursue higher-status careers than those found in the waste trades.[57]

Though the number of foreign-born participants in the industry declined with the end of mass migration, the participation of Jewish individuals remained high. Some were the children of immigrant founders who had found respectable livelihoods in growing family firms, even though many individuals in this category left the family business for other careers. The structure of the industry prior to 1945 allowed newcomers to the industry opportunities for upward mobility. Members of the second and third generations of immigrant Jewish families who had failed to ascend to the middle class found new opportunities dealing scrap due to demand generated by wartime shortages. Leonard Tanenbaum, the scrap dealer who observed racist abuse hurled at his employees, was a young, second-generation Jew raised in poverty (his parents briefly placed him in an orphanage) in Cleveland in the 1920s and 1930s. In the late 1930s, his father got him work in a factory that sorted discarded woolen clothes and processed them for resale to painters. The factory was owned by a Jewish entrepreneur and staffed primarily by Jewish workers who sorted and cut up old clothes. In the summer of 1941, he opened a scrap metal yard on Cleveland's east side to take advantage of shortages caused by the war effort and to perhaps find new, lucrative supplies to sell. His initial suppliers and customers were horse-and-buggy peddlers, many of whom rented their horses and wagons from livery stables.[58]

Tanenbaum rented a small shop where he could store what he bought. He developed relationships outside of the Cleveland metropolitan area, working with customers from New York, Chicago, Philadelphia, and other manufacturing centers who provided him with lists of items they wanted to buy. Tanenbaum tried to secure a few deposits from customers so he could have some operating capital to bid on the items they wanted.[59]

Tanenbaum's first break in the business came when he purchased a stack of silverware from dealers who had scavenged it out of a local dump. He made some inquiries and discovered that there was a market for reclaimed silverware. Manufacture of new silver had ceased and restaurants went out of business for lack of silverware. He learned from one of his buyers, a dealer in Chicago, that metal plating companies could make old, tarnished silver look like new through soaking it in cyanide and then tumbling it in a barrel, a process known as burnishing. Tanenbaum began purchasing and collecting old silverware, taking it to the plating company for treatment, and selling it to restaurants and hotels. Demand was sufficient enough that he soon exhausted local supplies and was traveling between thirty and sixty miles a day to find new sources. Tanenbaum expanded his business after the war, operating two scrap yards in Cleveland and passing them on to his children in the 1980s.[60]

Tanenbaum's tale of upward mobility from immigrant entry in the business to yard owner followed the common path forged since the late nineteenth century, yet it did so in a context where government regulations and an industry featuring established, decades-old firms had reshaped the market. Opportunities for further advancement from yard owner or processor to the level of scrap broker were less apparent, and the notion of immigrant businessmen working from rags to riches was far less common than it had been during World War I.

The economic context newcomers encountered allowed for less opportunity for advancement in the scrap industry than was available to earlier entrepreneurs. Firms founded twenty to sixty years before World War II had established shares of the market. The larger firms specializing in scrap processing had invested millions of dollars in technology to increase their volume traded. Moreover, large customers such as Bethlehem Steel renewed their attempts to collect scrap without the aid of independent dealers. Many manufacturers developed on-site yards and participated in scrap drives during the war that made them far more active in reclaiming industrial materials than they had been in 1940. Between the manufacturers and the large independent dealers, the move to larger, more complex operations meant that a novice to the scrap industry in 1945 was more likely to enter an established firm as a yard worker or—depending on his level of education—a lawyer or a metallurgist than he was to start his own peddling operation and move into owning his own yard. Scrap was important during World War II, and scrap operations grew big and complex.

Temporary Virtue

Between 1929 and 1945 major economic, political, and cultural shifts in the nation reshaped the scrap industry. Scrap firms were grudgingly respected at the beginning of the Depression as businesses engaging in a necessary practice, but one that was unpleasant and ran the risk of negative physical and moral effects on society. Moreover, scrap firms were about the only businesses in American society engaging in material reuse. Engineers in several industries attempted to manage waste in the name of efficiency, but their activities were confined inside factories and laboratories and out of sight of the general public. What material reuse was conspicuous involved scavengers and junkmen sifting through trash and piling scrap high in blighted eyesores of yards.

As scores of smaller firms were weeded out during the lean years of the Depression, the efficiencies of wartime production encouraged a rise in the power and market share of large corporations in many industries. The war certainly bolstered consolidation in the automobile and steel industries, and to a degree it led to consolidation in the scrap iron and steel industry.

The behavior of federal agencies regulating materials favored efficient allocation of resources to the largest businesses, establishing price controls and freight rates in an effort to gain control over scrap markets for the war effort. The OPA encouraged direct dealing of scrap from industrial suppliers to steel mills, often bypassing scrap dealers altogether. Other steel firms followed Bethlehem Steel's lead and opened their own yards to process and sort scrap. Many dealers went out of business during the war, and trade association officials voiced concerns that steel mills used the war effort to increase direct dealing and gain control over the scrap market.[61]

A simple model of decline for independent dealers during the war is not accurate. While many yards closed, still more dealers—including Leonard Tanenbaum—opened their businesses. Independent scrap dealers were not put out of business by steel mills during World War II, though their standing was challenged by consolidation of large segments of the American economy and by new, complex regulations.[62]

Scrap dealers large and small participated, as did steel and other manufacturing industries, schools, charities, voluntary organizations, and individuals across the nation in consumer scrap drives to aid the war effort. There is some debate over how effective the scrap drives were in finding resources for war industries, but there is no debate that Americans temporarily changed their attitude about waste materials. The key word is "temporarily." Scrap became a moral crusade during the war, but that crusade ended with V-J Day. Once the war was over, Americans abandoned their salvaging ways and embarked upon the most conspicuous period of consumerism the nation had ever seen.

5 | Size Matters

A festive mood was in the air in Michigan on 14 July 1966. On this day, Carl S. Albon, president of the Ogden Corporation subsidiary Luria Brothers, and Ben D. Mills, vice president of purchasing for the Ford Motor Company, led a group of men from both firms in turning over the first spadeful of earth at the two companies' joint venture in Taylor Township in suburban Detroit. Ford agreed to let Luria Brothers build a $3.5 million automobile fragmentation plant on the site, as well as a second plant in Cleveland. Luria had built a prototype of these plants in Los Angeles in 1963, giving it three fragmentizers that could handle large volumes of automobiles. Once in operation, the plants would chew up a quarter of a million junked cars each year, turning each vehicle into one thousand pounds of high quality steel scrap. The primary customer for this scrap, as negotiated in a long-term agreement, would be the Ford Motor Company. Albon and Mills enthused that the new fragmentizers would not only supply Ford with a significant source of high quality scrap but also, as Mills put it, remove over half a million "junked auto eyesores" from the Detroit and Cleveland scene each year.[1]

Such a project would have been unimaginable just two decades earlier. No scrap firm could have invested that kind of capital into processing technology. Ford might have, and, indeed, River Rouge saw many reclamation projects under Henry Ford's watchful eye, but why would Ford choose to enter into an agreement with an independent company rather than develop a plant on its own? And what had caused Luria Brothers to become a subsidiary of Ogden? It had been an independent, family-owned industry giant for over half a century. According to Albon and Mills, a principle benefit of the multimillion-dollar investment was the removal of junked automobiles from public view.

More, More, More

If World War II on the home front was defined by belt-tightening, the postwar period was defined as an era when bigger was better. The scrap industry grew large, in part because of economic and technological changes in both the scrap trade and related industries and because Americans were disposing of far more reclaimable scrap than ever before.

Rising standards of living over the next two decades meant more people spent more money on a variety of amenities. National output of goods and services doubled between 1946 and 1956, with private consumption accounting for two-thirds of the gross national product. Housing construction, automobile sales, amenities such as fast food, air conditioners, vacuums, and dishwashers proliferated in the suburbanizing nation.[2] Wartime austerity measures were abandoned. In *The Bulldozer in the Countryside* Adam Rome observes that consumers in the 1950s and 1960s were unable or unwilling to conserve energy as more houses were wired for electricity, and more people lived in the suburbs and had to commute to work in cities. Americans consumed more conspicuously than ever before.[3]

The corporations that serviced Americans' desire for housing, energy, automobiles, food, and other amenities were significantly larger in the 1950s than they had been before the war. While the scrap industry remained unusual in that it was a collection of several relatively small sellers catering to a few large customers, growth and consolidation were themes in scrap just as they were in the manufacturing industries. The largest scrap firms in the 1960s were much larger than their wartime counterparts, and due to a variety of expenses, a greater number of firms—including the firms handling the majority of volume traded—were owned by corporations rather than individuals or families. The corporations often retained family members as executives after sales, prizing their personal knowledge and contacts in the industry.

The era of bigger-is-better reshaped material reuse. Scrap collecting reverted back to private industry after the war, but the firms handling scrap were larger and in many ways different from those working in the 1930s. Furthermore, though the postwar period marked the end of federal price controls, changing patterns of consumption and aesthetics prompted the federal government to pursue a different method of regulation of the scrap industry in the 1960s.

Consumption and waste became conspicuous to the point that essayist John A. Kouwenhoven titled his collection of articles on life in America *The Beer Can by the Highway*. Contemporary readers understood the centrality of Kouwenhoven's image to their lives. In the years after World War II human pressures on the environment became dramatically visible, especially in matters relating to buildings, transportation, and energy. All were closely related aspects of the mounting environmental pressures of the time. Rapid construction of homes, commercial establishments, factories, shopping malls, and recreational centers was an ever-present feature of the postwar years. Growth ate up undeveloped land on the nation's urban peripheries as America became a suburban nation, residing, shopping, and conducting business away from city centers.

The sprawl of development came in tandem with expanded transportation. Americans increased their use of automobiles and airplanes to get from city to city. Innovations in transportation, including the development of the interstate highway system, took place after World War II to accommodate the increasing use of passenger automobiles, trucks, and airplanes. These vehicles required ever-greater amounts of fuel to run, and petroleum consumption escalated. Automobiles grew to new heights of popularity. As Americans purchased automobiles at unprecedented rates, they disposed of their old vehicles as they never had before. An estimated 25,000 automobile graveyards were scattered across the nation in 1951, and over eight million obsolete automobiles lay waiting to be scrapped in the mid-1960s, most in wreckage yards.[4]

The people charged with handling all of these disposed-of materials were in many ways the same people who had handled society's wastes before the war. Scrap collecting reverted almost immediately to the private scrap businesses that had been the source of so much prewar concern about hygiene and ethics, and the notion of all citizens collecting their old goods evaporated. The work remained marginalized in inner-city slums and on the urban periphery.

Perhaps the most significant change in the postwar scrap trade is that a few firms became much larger and more powerful than any in the trade's history. Brokers pursued import and export streams that had been interrupted by the war with new vigor. The geographic spread of manufacturing industries and consumers' use of items of mass consumption across the nation caused a spread of the scrap industry as well, though the major centers of exchange continued to be dominant. The consolidation of big business evident in the automobile, steel, and other manufacturing industries in the fifteen years after the war was felt in the scrap industry as well.

The regulatory environment of the immediate postwar period relaxed. Federal regulations of the scrap market, already eased as supply conditions were deemed less critical in the last two years of the war, became less conspicuous burdens on dealers. OPA price ceilings continued into November of 1946. A steel strike early in the year retarded demand, but upon the strike's settlement in February, consumption of scrap increased. The OPA alleged dealers hoarded scrap in the fall of 1946, though ISIS, citing depleted dealers' inventories, contended those allegations were groundless.[5]

Though the wartime trade brought many restrictions to dealers, dealers were in much better shape at the end of the war than they had been during the Depression. Almost four times as many firms operated in 1948 than operated in 1933, and the largest firms experienced unprecedented growth. Dealers flourished as federal regulations of the scrap markets eased in the postwar era.[6]

The public movement for reclamation of materials ended and reclamation reverted back to the private scrap dealers. Steel had been the defining scrap material of the twentieth century, shaping trade and individual firms to a greater extent than any other material.

Brand New Cadillac

The dominant symbol of postwar consumption was made largely of steel. Americans were encouraged to purchase new automobiles every year to keep up with the latest style. Automobiles also made possible new or expanded forms of consumption, from the drive-in restaurant to driving to the new suburban malls that opened for business in the 1950s.

The automobile provided the scrap iron and steel industry with both a continuing supply of scrap and a source of demand. It had been a major component of the American scrap market since its proliferation as an item of mass consumption after World War I. Production of passenger cars in the United States escalated from a few thousand at the turn of the century to over a million per year by 1915, largely due to Henry Ford's assembly line.

Competition from General Motors (GM) and smaller manufacturers pushed annual production to over three million cars per year by 1923. The Depression hurt production, with the nadir coming in 1932 when only 1.1 million passenger cars were made. Even with that dip in production and a cessation of production during World War II, the American automobile industry regularly churned out millions of new cars every year.

The war spurred consolidation of the industry, making the Big Three manufacturers among the world's largest corporations. They grew so big because automobiles became necessities in postwar American life. As Americans flocked to new homes in the suburbs, they relied on their cars to get to work, stores, school, and places of entertainment. Automobiles supplanted trains as the federal highway system grew in the 1950s.

Consumers viewed automobiles not simply as necessities but also as stylish status symbols; many bought a new model every other year. Between 1949 and 1965, Detroit regularly produced over five million cars per year; between 1965 and 1975 production escalated to between seven and nine million cars per year.[7]

The rise in automobile consumption in the United States brought with it a concomitant rise in consumption. The industry operated using the concept of planned obsolescence, building annual models that encouraged purchase of new vehicles as older ones both went out of style and became less reliable due to wear and tear. GM was the first American automobile manufacturer to introduce annual models; both GM and Chrysler used the strategy to quickly rival the Ford Motor Company in market share during the 1920s. Issues of style and durability encouraged consumers to purchase new vehicles every few years, or even annually.[8]

If the automobile was a defining characteristic of postwar American society, it was also a defining characteristic of changes in American material reuse—or, in many cases, a lack of reuse. Once OPA production controls were lifted, automobile production quickly became a dominant market for iron and steel. Consumers ceased collecting and returning items to the cycle of production and embarked on a new and unprecedented era of consumption and disposal, burning fuel in their new cars and disposing of those cars in a matter of months. Americans threw cars away by the millions.

Where did the old automobiles go? Some owners abandoned their cars in remote areas. Others sold them to specialty yards dealing exclusively in junked vehicles. These yards, common to the edges of American cities since the late 1920s, proliferated in the 1950s, frequently at the side of highways. The contents of automobile graveyards included potentially cheap replacement parts for repairs, or grist for scrap processors. In times of low demand, they produced highly visible piles of rusted car bodies for passersby to see.[9]

Automobile graveyards, usually located alongside major roadways on the outskirts of cities, grew in size and number over the next several decades. Prior to the United States' entry into World War II, these yards were recognized as valuable sources of scrap metal for defense. The American Automobile Association estimated that the government could salvage five million tons of steel from the nation's automobile graveyards.[10] During a scrap shortage in 1951, the federal government mandated that automobile graveyard operators be required to turn over their inventory to mills every ninety days in the hopes of generating an extra one million tons of scrap sold annually. (At the time, automobile graveyards were the estimated source of about four million tons of scrap iron and steel sold that year.)[11]

America's love affair with the automobile spurred increased consumption of cars throughout the 1960s, making the automotive industry the nation's single largest market for iron and steel (the construction industry comprised the second largest market for those metals). By 1970, automobile manufacturers consumed 20 percent of all steel products and gray iron castings and 40 percent of malleable iron castings in the United States.[12] The automobile provided both demand for scrap and a major source of supply. Automobile production generated large amounts of prompt (internally generated) industrial scrap, accounting for perhaps 25 to 35 percent of such scrap sold in 1970. Automobile producers either processed this scrap themselves or contracted large processors such as Luria Brothers to handle it, usually with equipment located at the automotive plants. Such scrap was highly desired due to its known composition.[13]

Automobiles were wonderful containers of extractable steel, yet the challenge for dealers was extracting the steel from the glass, rubber, lead, plastics, and cloth embedded in the complex consumer product.

The complexity of automobiles—and other consumer products, including refrigerators, television sets, and air conditioners—prompted innovations in the scrap industry that helped change the size and operations of scrap firms.

Automobiles were not the only sources of abundant scrap after 1945. Large public works projects and industrial expansion produced vast quantities of construction waste, obsolete military machinery, postindustrial scrap, demolition wastes, and postconsumer discards, some of which were sold at auction or abandoned for scavenging. Obsolete military equipment had been a crucial source of salvage since the Civil War, and as the growing military-industrial complex of the cold war ensured sustained military production, the amounts of decommissioned materials rose after World War II.

Urban renewal efforts led to slum clearance in dozens of cities, resulting in the scrapping of structural steel, concrete, wood, lead, and other materials found in demolished buildings. Subsequent construction efforts also generated wastes, as did production in several industries. The net effect was a large increase in the variety and volume of industrial wastes, many of which were reused in-house or sold on the open market.

The Rising Cost of Business

What became of all these materials? Prior to World War II, most scrap firms were small, family-owned establishments that remained in the hands of their immigrant founders. The technology used in most yards was moderate in price and had not substantially changed since the turn of the century.

The two decades after the war would be among the most complicated for scrap dealers. Some reaped massive profits and others were forced out of business due to changing demand for materials, changing technologies, increased labor and equipment costs, and the changing structure of the industry. One of the more important dynamics affecting the fortunes of individual firms was the changing demand for purchased scrap by industrial customers. Innovations in industrial production altered demand for several scrap materials after 1945. The steel industry, long a proponent of the open-hearth process of forging steel, began adopting the Basic-Oxygen Steelmaking Process (BOSP) in the 1950s. The process, which used pure oxygen rather than air to convert steel, was less demanding about the type and quality of iron being refined than the open-hearth process was. Consequently, steel mills could use even larger percentages of scrap iron (and scrap iron with more impurities) by adopting the BOSP.

Developed in Austria in 1952, the process was quickly adopted in Europe. Jones and Laughlin Steel was an early American user of the BOSP, though many American mills chose to retain the open-hearth furnaces in which they had invested until the 1970s. International adoption of the BOSP was more rapid, and by 1970 it was the most widely used steelmaking process in the world.[14] The process allowed mills to use home scrap, and while scrap consumption increased in the 1950s and 1960s, the ratio of purchased scrap as part of the total scrap consumed fell. Industrial material reuse in the ferrous industry shifted from purchasing large quantities of postconsumer material to eliminating factory wastes through more efficient use of materials the mill already owned. American steel companies were slow to adapt to the new process, but its adoption by European companies increased the use of scrap iron in European mills during the 1950s, spurring exports from American scrap yards. The global scrap trade intensified in the 1950s with the formation of the European Coal and Steel Community in 1952, an organization that both promoted the scrap iron trade within Europe and facilitated trade with American brokers.

U.S. steel production increased its use of ferrous scrap, though shortages on the open market in 1951 and innovations in reclaiming home scrap meant that mills relied on their home scrap for production needs. The Battelle Institute conducted a study concluding that purchased scrap accounted for 42 percent of the total scrap consumed from 1950 to 1956, with the majority coming from home scrap generated within mills.[15]

Not all innovations increased demand for scrap materials. Scrap rubber was a highly sought commodity between the mid-nineteenth century and World War II due to rubber's relative scarcity and utility. As automobiles became items of mass consumption, scrap rubber was a critical ingredient in the tire industry, as no acceptable substitutes for rubber existed.

Demand for rubber or rubberlike substances during the war spurred the U.S. government to embark on a "Manhattan project" for synthetic rubber. This effort began a period in which synthetic rubber and plastics found widespread applications as affordable substitutes for natural rubber.[16] By the 1960s, plastics were common in packaging materials, medical instruments, toys, and almost every appliance, automobile, and complex consumer good on the market. Synthetics replicated the durable, flexible qualities of rubber in a material of limitless abundance that made consumer goods that had depended on rubber cheaper. Since rubber was no longer a prize commodity, demand for scrap rubber declined to the point where dealers were left with stagnating supplies for months or years at a time.[17]

Markets for ferrous scrap were transformed by changing technology in the steel industry. Electric-furnace and continuous casting steel production could use charges composed entirely of scrap iron and steel. Continuous casting, a process that reduced operating costs by bypassing many of the traditional stages of heating and cooling steel, was developed in Germany in the 1930s and adopted widely throughout the world in the 1950s. The initial process featured a vertical, open-ended, water-cooled mold, which oscillated to prevent the molten metal from adhering to its walls. Casting was continuous in the sense that the size of the ingot was not determined by the size of the mold; rather, given a steady supply of molten metal, the cast product would emerge like toothpaste from a tube. Steel manufacturers adopted mechanisms to rapidly cool the molten steel as it passed through and out of the mold. Continuous-casting facilities included cranes and ladle cars for handling molten steel. An intermediate reservoir strained out slag as the metal entered the mold. Water sprays chilled the steel after it left the mold, then rolls drew the steel from the mold. Once out of the mold, the cooled steel went before a saw or torch to cut the cast product into the requisite lengths.[18]

The innovations in steel production initially hurt scrap dealers. Mills began piping oxygen into open-hearth furnaces in a process that improved the furnaces' speed and efficiency. According to U.S. Steel, whereas many mills melted steel from a charge containing 40 percent to 60 percent scrap, newer furnaces used a charge of as low as 30 percent scrap by 1962. Mill purchases of scrap, affected in part by the new techniques that depressed demand for steel, shrank from 35.1 million tons in 1956 to 23.6 million tons in 1960.[19]

Although demand varied between 1945 and 1965, the complexion of the scrap trade during this period retained similarities with the prewar trade. Most activity remained centered in the mercantile and industrial centers of the Northeast and Midwest, with the states of New York, New Jersey, and Pennsylvania hosting the greatest volume of materials traded and the largest number of firms operating. Concentration of activity conformed to patterns of urban development and industrial activity. The largest population centers in 1948 featured both the largest number of what the census identified as waste material firms and the largest volumes of waste material sales.[20]

Metropolitan areas with over one million residences accounted for 45 percent of the nation's waste material firms and 53 percent of the nation's sales volume, with the industrial Midwest accounting for the majority of activity outside the New York–New Jersey–Pennsylvania region. Chicago's waste materials dealers had the second-greatest volume of sales in the nation after New York City. Philadelphia was a distant third, followed closely by Cleveland, Pittsburgh, and Detroit. Those six metropolitan areas were the only ones generating over $100 million each in annual sales volume.[21]

The 1948 Census of Business revealed one more continuity with the prewar scrap trade. Most scrap material dealers, brokers, and processors operating in the United States were small firms owned by individual proprietors or partnerships, with many remaining in the hands of the immigrants' families who founded the businesses in the late nineteenth and early twentieth centuries. The 1948 Census of Business enumerated 3,044 iron and steel scrap firms operating in the United States, with reported annual sales totaling $1,696,195,000. Almost one-third of those firms were small dealers doing an annual business of less than $50,000. These firms generated under 2 percent ($22,122,000) of the industry's annual sales. Concentration of the market is evident in the small number of firms generating over $5 million in sales. Only sixty such firms were reported, accounting for 2 percent of the total number of firms. This 2 percent generated 43 percent of the total annual sales for the industry.[22] Over half (1,581 of 3,044) of all scrap iron and steel firms enumerated in the 1948 census were classified as individual proprietorships. One-quarter (803) were listed as partnerships; 22 percent (663) were listed as corporations; and the remaining seven firms were classified as "other forms of organization."[23]

This continuity was short-lived as the dynamics of the trade led a few brokers at the upper end of the industry to concentrate their market share in the late 1940s and early 1950s. In 1949, *Fortune* reported "the bulk of the 'dealer' scrap still passes to the steel industry through about 100 brokers, some of whom also own yards." The magazine attributed this "funneling of a huge scrap tonnage" through the brokers to a need to simply and concentrate a potentially scattered and chaotic market. Experienced brokers could use their skills to facilitate scrap flows, and they were still valuable to mills despite the increase in direct dealing.

Managing Salvage

The most successful brokers in 1949 were second-to fourth-generation scrap men who grew up in the industry and relied upon both the skills they learned in the trade and the American managerial skills they learned in formal educational institutions. *Fortune* asserted that newcomers would find it extremely difficult to set up competing brokerages, as the established networks of business contacts and market expertise gave existing brokers a tremendous competitive advantage. Even direct dealing could not replace the brokers. Few steel companies were able to purchase scrap directly as effectively as they could with a broker; *Fortune* noted that Bethlehem Steel's yards were unusually well run, the result of two decades of tough campaigning.[24]

Brokerage firms varied in size, though assessing exact size is difficult. They were privately held, usually individual proprietorships or partnerships, and did not make their accounts public. Schiavone-Bonomo and the David J. Joseph Company were large brokers, certainly among the sixty firms generating $5 million in annual sales in 1948. Luria Brothers was universally regarded as the largest broker, believed in 1949 to do as much as one-seventh of the entire American scrap iron brokerage business. Luria Brothers was a large company relative to the rest of the industry in the first half of the twentieth century; the firm did an annual business of $100,000 in 1901; $1 million in 1909; and more than $100 million in 1949. Their holdings in 1949 included a Philadelphia headquarters, fourteen branch offices, six preparation yards, and a labor force totaling several hundred.[25]

The company's growth was headed by a management team that could be characterized as long serving and stable, with some family connections but also a record of reliance upon younger hires from outside the family who had some higher education. The team included the largely retired Alex Luria and Executive President Joel Claster, who was Alex Luria's son-in-law and the past president of ISIS. From outside the family, Robert H. Clymer, vice president in charge of yard operations, came to Luria Brothers as a fifteen-year-old office boy in 1907. By the late 1930s, Hirsch Luria's university-educated grandsons began entering the family business, and a decade later they made up much of its management. President Herbert B. Luria, Vice President Henry

T. Luria, and Secretary Mortimer Luria all graduated from Yale. Henry Luria subsequently earned a law degree from Harvard. Treasurer William F. Luria studied at Babson Institute, and Vice President David Luria attended Lehigh and Oxford.[26]

Vice President Amos Bowman, in charge of Pittsburgh brokerage, joined the firm in 1917. Vice President William H. Hundt, supervisor of the New York area, joined Luria Brothers in 1911. Vice President William J. Luria, Alex's nephew, started with the firm in 1929 and headed its Philadelphia brokerage in 1949. Vice President George Stout was Luria's chief ship breaker; he joined the firm in 1924. Vice President Herbert J. Biel, head of the Chicago office, joined Luria in 1936 after graduation from Syracuse University. Alex's son Herbert L. Luria joined the firm in 1933 and was treasurer in 1949. Secretary William L. Forebaugh joined Luria in 1920. In 1949, the youngest executive was thirty-two-year-old Ralph E. Ablon, an Ohio State University graduate and Navy veteran who joined the firm in 1940. His duties included directing import and export activities.

Luria's management represented a mix of family members and professionals brought in from the outside. Both groups consisted of individuals trained in business and law schools, bringing mainstream American corporate values and expertise back to the firm. The diversification of the firm's executive structure, however, did not reduce the emphasis on the unique skills of evaluating and handling scrap materials. Despite their white-collar educations, all management recruits were compelled to begin their careers with at least six months' work in the scrap yards handling and evaluating materials so that they could gain the skills that built the company.

Luria Brothers' growth brought with it the formation of new companies spun off from the original firm. Luria Steel and Trading Corporation was established in 1935 when Alex Luria's older brother Max decided to concentrate on exporting finished steel and scrap. Max Luria kept just under half ownership in Luria Brothers, and Alex Luria held about the same portion of Luria Steel and Trading. Max Luria died in 1939. When his five sons returned to the firm from war service, the cross-ownership with Luria Brothers was terminated by mutual sale, and Max's sons plunged into the domestic scrap-dealing business.

Luria Steel and Trading grew rapidly. By 1949, the firm had absorbed two other brokerages and operated offices in eleven cities, including New York City and Little Rock, as well as a large yard in

Chicago specializing in railroad equipment. The firm's chairman of the board was Joseph Rovensky, an old family friend and a vice president of Chase National Bank.[27]

New Management, New Technologies

Luria's growth and spread across the country was extraordinary, yet aspects of that growth were representative of changes occurring within the industry. As firm founders were succeeded by younger generations, the new leaders often had educations as trained lawyers, financiers, and metallurgists. The structures of companies featured greater specialization, with legal departments, more extensive accounting procedures, import/export offices, and engineering divisions. Investments in technology increased, allowing firms to process greater volumes of material with greater speed than before. Scrap shears continued to grow in size and automation. Investment costs increased as the use of processing technology, labor, and transportation allowed firms to process and deal greater volumes of material with greater speed than before. The conservative implementation of new technology practiced in the 1930s gave way to a period in which firms recognized new investments were necessary if they wished to compete. Those that did not—or could not—invest were at a competitive disadvantage.

In the immediate postwar period, investments in technology remained conservative. The *Waste Trade Journal* surveyed four thousand dealers of scrap metal, rubber, waste paper, and rags in 1948 to assess the technological requirements of waste material dealers. The survey found that waste paper and rag dealers reported anticipated purchases of many new items. The most popular items desired were bale stackers, bale ties, baling presses, bins, box trucks, conveyors, cranes, dust collecting systems, fork trucks, fluorescent lighting, grease and oil, laundry equipment including washers, centrifugal extractors and drying tumblers, hand trucks, lift trucks, rag cutting machinery, rag shredders, trailers, and trucks. Published investment budgets ranged from $21,000 in replacement equipment to $85,000 to build a new waste paper plant.[28] Scrap rubber dealers reported anticipated purchases of cranes, pallets, scales, bead cutters, and trucks, with the largest published purchases being $10,000 for pallets and $15,000 for a crane.[29]

The scrap metal industry reported more extensive investments in equipment than did other waste material dealers. Common purchases by scrap metal dealers included hydraulic press equipment, cranes, shears, dump trucks, fork trucks, and furnaces. The largest annual investment reported was for $230,000 in equipment, including $80,000 for baling press equipment, $50,000 for overhead cranes, and $40,000 for tractor cranes. Smaller scrap metal dealers reported anticipated investments ranging from $200 to $175,000.[30]

The M. Cohen and Sons Company plant in Cleveland represented one example of how processors employed their equipment in 1948. The twelve-acre yard featured one and a half miles of rail track organized for a constant flow of material from left to right. All points in the yard that were accessible to automobiles were also accessible by the six rail tracks, maximizing yard workers' ability to move carloads of scrap.

Scrap coming into the yard went onto a 40' x 10' scale with a 100,000-pound capacity. Once a carload of scrap was weighed, the routing foreman routed it toward a baling press if it was light No. 2 scrap, toward shears or a skull cracker if it was heavy No. 1 scrap, or elsewhere if it was specialty material. Heavy scrap went to one of the four Canton direct drive automatic shears, including one No. 5 shear suitable for thick steel such as that from ship hulls and three smaller No. 3 shears suitable to cut lighter grades. Crews of three to five men fed steel into the shears. Two overhead cranes moved the sheared scrap to a warehouse. Large, unwieldy pieces of scrap that could not be easily transported to or cut by the shears were dealt with in a primitive but effective manner that scrap yards had used since World War I. At the back of the yard, a skull cracker (an eight-thousand-pound ball) was dropped on large pieces of heavy cast iron to smash them into smaller pieces. The area in which skull-cracking took place was securely walled off from the rest of the yard to avoid injury and damage to equipment from large pieces of flying metal.

Near the front of the yard, gondola cars loaded with loose scrap iron were guided by one of four locomotive cranes to a compression pit. One operator guided the crane and another controlled the hydraulic compression machine (or baling press) to press the loose scrap into dense, cubed bundles. The bundles were made to weight specifications by steel mills, ranging from 500 to 1,200 pounds apiece. Bundles fed out of the press onto a conveyor belt and into gondola cars.

Not all scrap processed in the yard went through the large shears and balers. Yard workers cut scrap with seventeen oxy-acetylene torches, just as their predecessors had for decades. Once scrap metal had been cut by torches and shears, smashed by skull crackers, and baled in the compression pits, gondola cars transported it to the company's on-site 220' x 160' warehouse. There it remained until sold to steel mills.[31]

M. Cohen and Sons' investment in technology was greater than that of earlier dealers, though the equipment purchased—shears, skull crackers, rail tracks, gondola cars, baling presses—resembled the kinds of machines employed in scrap yards since the turn of the century. Over the next fifteen years, scrap metal dealers' technological investments increased, incorporating not only variations on existing technology but also applications of equipment new to the industry. The size and capacity of shears and balers increased substantially. New processing machinery, including the fragmentizer, the hydraulic shear, the hot briquetter, the motor breaker, and the rail breaker proliferated between 1945 and 1960.[32]

Hydraulic shears could cut six-inch-thick girders down to sizes that could be fed into a steel furnace. These shears, which could cost over $100,000 to purchase, were large and complex enough to require a crew of three to five men. They could produce approximately three times the volume what the same number of men using other shears could produce.[33]

With the automobile rising as a source of scrap, however, acetylene torches and shears proved inadequate to quickly separate steel from the rest of an automobile. Alton Newell, a scrap dealer working in New Mexico, developed a technological solution to this problem. Beginning in the late 1930s, Newell began work on machinery designed to harvest scrap metal from automobiles by hammering and shredding the automobile and using magnets to separate the ferrous scrap from other materials. By the late 1950s, Newell's automobile shredder had become a staple of the scrap yard as its ability to hammer down an automobile and quickly separate light iron and steel from the many other materials found in automobiles gave operators the ability to process a massive source of consumer-generated scrap. A scrap processor could take the husk of a car (usually scrap yards acquired automobiles from auto parts dealers after those dealers had removed the tires, fuel tanks, batteries, radiators, and valuable components) and feed it into the shredder. A series of hammers within the shredder fragmentizes the automobile into pieces of metal about the size of a fist.[34]

This machinery was more expensive than its predecessors, and many older firms could not afford the shredder. Those that did not adopt the new technology were at a competitive disadvantage to the firms that chose to shred. A new complication of using automobile

Figure 5. Dempster Dumpster system, 1966. Processing junked automobiles efficiently required investment in new, expensive technologies. *Courtesy Hagley Museum and Library.*

shredders was the concern over contamination of extracted iron. Dirt and sulfur-laced rubber increased the amount of slag produced when scrap was used in steelmaking, slowing the process down. Oil and grease could clog vents and increase temperatures in the open-hearth furnace, disrupting the process. Lead and aluminum could damage the furnace.[35]

Magnets could separate shredded ferrous scrap from other metals, and oil, grease, and dirt could be washed away. Magnets left copper to contaminate and harden steel, though processors could minimize copper's effects by adding nickel. Mill supervisor R. F. Kuhnlein stressed that separation of materials was crucial before scrap was bundled. "All of the undesirable items not wanted in scrap can be found in hydraulic bundles. Inspection is difficult because of the skill with which these materials can be hidden on the inside of a bundle."[36]

Shredders and balers produced new problems for scrap dealers and customers, but they offered benefits that outweighed the concerns. Shredders chopped automobiles and appliances into fist-sized chunks.

Balers could crush large loads of loosely bundled scrap into twelve-cubic-foot blocks. These small, densely packed bundles of metal lowered the costs of handling and transportation, thus lowering prices for customers and increasing profit margins for dealers.[37]

Changing transportation costs caused changes to the methods of transportation scrap dealers used. Scrap remained a rail-oriented commodity in the United States after World War II. Major steel mills were set up to handle rail and, to a smaller degree, water shipments when their plant was on a navigable body of water. Railroads throughout this period maintained large fleets of gondola cars suitable for moving scrap, steel, and other bulk commodities.

Most scrap processors continued to receive large amounts of material from local sources, such as automobile wrecking yards, small machine shops, stamping plants, farmers, and scavengers. Many scrap preparation yards removed scrap from industrial sources using dump trucks, flatbed trucks, and specialized roll-off containers. These containers could carry as much as twenty tons of dense material in gondolas on railroads. Many scrap companies purchased used open-top trailers, which they stored at suppliers' plants until the supplier filled the trailer.

Scrap firms with access to navigable waterways enjoyed cheaper transportation costs. Exporters on the coasts and major rivers developed large accumulation facilities by 1960. Some firms purchased their own docks and cranes, enabling them to load vessels with capacities of 15,000–30,000 tons. Some of the large export scrap producers purchased their own ships or chartered vessels to move material to the scrap-consuming countries of Europe and Asia.[38]

Changes in processing technology and transportation increased expenses for scrap processors and contributed to rising labor costs. Large trends in American industry affected the scrap trade. The CIO began organizing large scrap yards' labor in the 1930s; by the 1950s, unionized workforces were common in most mid- to large-sized firms. ISIS members complained about the threats of work stoppages, and trade literature stressed the savings inherent in operating safe work environments, especially ones where unions had not organized the workforce.[39]

Workplace safety had long been an issue for the scrap industry. Companies sought insurance for their workers as early as World War I, and some suits under state worker's compensation were brought during that time. As scrap yards grew larger and handling capacity grew, the issue gained visibility. By the mid-1960s, ISIS provided recommendations to its members on purchasing affordable workmen's compensation insurance and maximizing yard safety.[40] In an industry where the handling of jagged, heavy, sometimes rusty materials and the use of increasingly powerful processing technologies were common, workplace safety was an ever-present and increasingly expensive issue. Furthermore, contamination of scrap by dangerous materials could imperil workers. In 1949, Nathan Trottner, a yard owner in San Antonio, was killed when one of his employees used an acetylene torch to cut a pipe that contained a small amount of nitroglycerin.[41] As scrap dealer Edward Fields noted, "Safety pays off in ways other than low [insurance] rates. Men who are able to walk and use both hands usually work better than the cripples we have all seen around some yards."[42]

Fields may not have used the most humanitarian rhetoric in advocating workplace safety, but his blunt point was well taken by his peers. In 1961, an insurance executive speaking at an ISIS convention reported that scrap metal workers suffered more disabling injuries per one million man hours than steel mill workers, coal miners, or foundry workers.[43] Common injuries included cuts, punctures, injuries due to shears and handling of other machinery, eye injuries, back strains, and hernias. The need for affordable workmen's compensation made workplace safety an issue in scrap yards, leading firms to train laborers to use equipment and invest in equipment that would reduce the risk of injury to its operators.[44]

Scrap Goes Corporate

The costs of operating large scrap firms were often too great for family-owned firms to sustain. Luria Brothers' response to the costs associated with its explosive growth provided a solution by example. In October of 1955, the Lurias sold the family firm to the Ogden Corporation, a holding company with diversified interests, for $20 million. New ownership did not eliminate the role of the veteran scrap dealer within the firm. Ogden, recognizing that much of the firm's value rested in the expertise and business relationships the Luria family had cultivated, retained Luria's executives after the sale.[45]

Luria's position as the dominant scrap firm was unique, but its transition to corporate ownership was not unusual. According to data recorded by the Census of Business, a shift in the type of ownership commonly found in scrap firms occurred between 1948 and 1963. The

1963 Census of Business enumerated 4,065 iron and steel scrap firms operating in the United States with reported annual sales totaling $2,066,083,000. Almost half (1,921 of 4,065) of all scrap iron and steel firms enumerated in the 1963 census were classified as individual proprietorships; 675 were listed as partnerships. Over one-third (1,450 of 4,065) of scrap firms were listed as corporations, up from the 22 percent enumerated in 1948. The remaining nineteen firms were classified as cooperative associations or "other forms of organization." A clear trend toward corporate ownership in the scrap industry was forged between 1948 and 1963.[46]

Firm structure dictated the level of investment in technology and yard innovations. Many scrap yards were able to afford new shears, shredders, and balers, but Luria Brothers, aided by its early transition to corporate ownership, purchased equipment that extended its competitive advantage over the rest of the industry. Armed with the capital available from corporate ownership and income from exclusive contracts with large customers, the company contracted Logemann Brothers in 1965 to build the world's first fully automatic scrap baler, capable of baling over 2,500 tons of scrap iron per day. Luria Brothers installed the baler, which weighed over a million pounds, at Bethlehem Steel's Burns Harbor, Indiana, plant, where it had exclusive rights to Bethlehem's home scrap at the facility.[47] One year later, Luria Brothers joined with Ford Motor Company to build the auto fragmentizers that would supply Ford with hundreds of thousands of tons of ferrous scrap each year. These investments were unique to Luria, resulting from its large operating budget and exclusive contracts.

Luria Brothers' embrace of corporate ownership was motivated by not only growth but also unwanted interest from the federal government. The firm's history in the postwar period reflected two ways in which the U.S. government took an interest in the nation's material reuse between 1945 and 1965. One way involved trade practices; the other, aesthetics. The trade concerns were specific to Luria Brothers' growth. At the same time the company began expanding its operations it also entered a protracted legal battle with the U.S. government, a battle that—though it was unusual—indicated how significantly the markets in postconsumer and postindustrial scrap materials had changed since 1930. A protracted antitrust investigation of Luria Brothers initiated by the Federal Trade Commission in 1953 mushroomed into a Senate investigation that concluded Luria's sales totaled $500 million by 1959, including half of the United States' export scrap business and nearly 40 percent of the domestic scrap business. The firm's postwar growth was explosive, with sales increasing over 400 percent between the end of World War II and 1959, and federal concern over the growth no doubt contributed to the founders' decision to cash out.[48]

The FTC charged that Luria used a wide range of practices to monopolize the industry. Luria stood accused of making loans to and purchases of competitors. The FTC claimed Luria raided its competitors' personnel and enjoyed "complete domination of the major scrap trade association [ISIS]." Finally, the FTC charged that Luria installed multimillion dollar superyards physically located on steel mill property and bound to the mill by exclusive dealing contracts.[49] According to a report by the U.S. Senate Select Committee on Small Business, Luria's power to control and fix prices, exclude newcomers, eliminate competitors from regional, national, and foreign markets, and force competitors to sell to Luria if they wished to remain in business all contributed to its effective monopoly.[50]

The Senate committee concluded that the FTC case against Luria was not sufficient to eliminate the firm's monopoly control over the industry and that criminal proceedings under Section 2 of the Sherman Act, aimed at breaking up Luria's national and regional monopolies, were necessary. These proceedings were unsuccessful, but the FTC case continued, threatening to curb Luria's expansion and impose penalties.[51]

The firm took the case seriously. In its defense, Luria Brothers provided several witnesses who vouched for the company's unique ability to serve its customers. Customers testified that they dealt with Luria Brothers due to the company's reliability. Charles Tyson, executive vice president of John A. Roebling's Sons Corporation, a subsidiary of the Colorado Fuel and Iron Company, testified that Luria provided the best quality of materials and was able to supply the quantity needed at a satisfactory price.[52]

Customer satisfaction was not an effective defense. In 1961, FTC examiner John Lewis ruled that Luria Brothers entered into exclusive supply agreements that were monopolistic and illegal. In reaching his conclusions, Lewis outlined many of the major shifts in the industry between 1945 and 1960. He concluded that while Luria had always been a large broker, its anticompetitive growth occurred after World War II. Luria was a substantial factor in the scrap business by 1945, but its influence was limited largely to the eastern United States. With the help of exclusive arrangements with large mills, the company became dominant in other sections of the nation by the mid-1950s. In 1947, Luria supplied about 17 percent of the scrap purchased from brokers and dealers by major U.S. steel mills; the firm's share of this market more than doubled by 1954, due (according to Lewis) to the exclusive sales agreements. Luria's share of the scrap purchased by mills with which it had entered into such agreements increased from 35 percent in 1947 to 78 percent in 1954. "Aided in large measure by the exclusive arrangements with the respondent mills," Lewis concluded, "Luria has become the most important single factor in the scrap industry in the United States, and the only company which is truly national in scope."[53]

Lewis ruled that Luria's acquisition of former competitors violated the antimerger section of the Clayton Antitrust Act by substantially lessening competition. He recommended forcing Luria Brothers to sell two Pittsburgh-based subsidiaries, the Pueblo Compressed Steel Corporation, which it had acquired in 1946, and Southwest Steel Corporation, acquired in 1950. Both had been rival scrap firms. Lewis pointed out that Pueblo was one of the largest dealers in the Rocky Mountain area at the time its stock was acquired by Luria in 1946, and its purchase "was of material aid to Luria in carrying out its exclusive arrangement with [Colorado Fuel and Iron's] Pueblo plant and in becoming the dominant factor in the market." Southwest was the second largest supplier in the Pittsburgh-Youngstown area when Luria purchased it in 1950. "Its acquisition," Lewis declared, "has materially aided Luria, which was already the largest single factor in the market, in far outdistancing all its other competitors in the market." As a result of Southwest's purchase, Luria's share of the Pittsburgh-Youngstown scrap supply market rose from 26 percent in 1949 to 36 percent in 1954.

Luria's anticompetitive practices were not limited to the domestic scrap market. At a time when European consumption of scrap in its new BOSP mills made American exports lucrative, Luria's behavior in the export market helped consolidate its dominance. Lewis ruled that Luria was the dominant one of three large brokers (the other two being Schiavone-Bonomo of Jersey City and Western Steel International Corporation of New York) supplying scrap to European markets. The three exporters as a group had an exclusive-dealing understanding with the Office Commun des Consommateurs de Ferraille (OCCF), the buying agent for European steel mills affiliated with the European Coal and Steel Community. Lewis concluded that the group supplied between

90 and 95 percent of all American scrap shipped to OCCF countries between 1954 and 1955. He ordered an end to these arrangements.

Lewis did not sustain all of the FTC's charges against Luria Brothers. He dismissed charges that the firm's acquisition of five companies—A. M. Wood and Co., Inc., Philadelphia; Lipsett, Inc., New York; Lipsett Steel Products, Inc., Brooklyn; Apex Steel and Supply Company, Chicago; and Cermack-Laflin Corporation, Chicago—in the 1940s violated the Clayton Act's antimerger provisions. He dismissed the charge that Luria and the Hugo Neu Corporation of New York restrained trade unlawfully by combining to act as the exclusive supplier of scrap for five Japanese steel producers. Lewis also dismissed allegations that seventeen American steel mills conspired to use Luria as their exclusive broker and that they conspired with Luria in several ways. The dismissed allegations included coercing railroads to sell scrap to Luria, selling new steel on the condition that resulting scrap would be sold to Luria, purchasing scrap at prices so high that it could only be resold at a loss, and making loans and advances to scrap dealers on the condition that they sell their scrap to Luria.[54]

In December 1962, the FTC upheld Lewis's orders to end Luria's monopolistic practices. It noted that between 1947 and 1954 Luria's percentage of the total scrap sales by all brokers and dealers in the United States rose from 17.1 percent to 33.7 percent, in large part due to exclusive agreements. The FTC ordered a dozen domestic steel producers—including Bethlehem Steel and its subsidiaries, U.S. Steel, National Steel, the Colorado Fuel and Iron Company, and Lukens Steel—to purchase over half of their annual scrap requirements from brokers other than Luria for five years.[55]

Luria appealed the case in the federal courts with little success. In 1968, the Supreme Court upheld the Federal Trade Commission's order barring Luria's exclusive agreements, putting into effect the FTC's command that Luria discontinue its exclusive arrangements with a dozen major scrap consumers. The decision also barred Luria from supplying more than half the scrap used by those consumers and barred Luria from buying up any other scrap firms for five years.[56]

Despite the sanctions levied on the company, Ogden's Luria Brothers subsidy remained the largest scrap firm in the world, albeit by less of a margin, over brokers such as David J. Joseph and Schiavone-Bonomo than it had enjoyed in 1955. It continued to do business with the world's largest scrap consumers, investing in millions of dollars of processing and transportation equipment to maintain its standing in the industry.

Neon, Junk, and Ruined Landscape

In addition to expressing sufficient concern over Luria Brothers' dominance of the ferrous scrap industry to intervene and prosecute, the federal government also began to take an interest in regulating the aesthetics of old material storage and handling. Ford and Luria's agreement to open the giant fragmentizers in Detroit and Cleveland continued the impulse toward consolidation in the scrap industry. The two firms' insistence that the new plants would remove a quarter of a million eyesores from the Detroit and Cleveland scene each year indicated what had become by 1966 the national impulse to control the aesthetic blights of automobile graveyards. The scrap industry had for decades experienced a long history of state and municipal government regulation. Progressive legislation after World War I pushed scrap yards out of many urban residential neighborhoods and into the unpopulated urban periphery and poorer neighborhoods lacking political power to oppose their siting.

The Highway Beautification Act of 1965, however, was the first federal law regulating the scrap industry that did so on aesthetic grounds. The battle over highway beautification took place at a time when the American highway system developed concurrent to unprecedented levels of mass consumption in American society. Consumption of land for residential housing increased as millions moved to urban peripheries. As Americans purchased automobiles at unprecedented rates, they disposed of their old vehicles as they never had before. By 1951, an estimated twenty-five thousand automobile graveyards were scattered across the nation.[57] By the mid-1960s, over eight million obsolete automobiles lay waiting to be scrapped, most in wreckage yards.[58]

Zoning ordinances after World War I pushed many yards to the urban periphery, where low land values, lack of neighbors, and quick access to rail, trucking, or shipping lines suited yards. As Americans migrated from central cities to suburbs in the 1950s and 1960s, exurban scrap yards once isolated from public view saw new residences, businesses, and roads develop around them. Interstate highways brought new levels of traffic past the previously isolated yards.[59]

Two problems emerged with the proliferation of salvage yards. One, they were often located near major roadways, so their activities were prominent. Two, demand for scrap metal was volatile. In periods where demand for scrap iron was depressed, as it was at the end of the 1950s, automobile husks piled high in yards. Complaints about automobile graveyards multiplied in the early 1960s, prompting new political action to reduce the nuisances associated with them.

Concern over the proliferation of junkyards and billboards by the side of the nation's highways grew in the early 1960s. *Reader's Digest* and other periodicals called for laws to regulate the nuisances. Lady Bird Johnson became the focal point for political action on the issue. By late 1964, the first lady had become a vocal advocate for removing billboards and junkyards from America's roadsides. She sought to beautify the highways, calling for "pleasing vistas and attractive roadside scenes to replace endless corridors walled in by neon, junk, and ruined landscape."[60]

Lyndon Johnson coordinated his staff to develop legislation to control aesthetic blights. He established a Task Force on Natural Beauty, with Lady Bird as its direct liaison to the president, to develop specific policy recommendations on the subject. In late 1964, the task force presented a set of recommendations to the president urging him to establish federal regulations of billboards and junkyards. A few days after the 1964 election, Johnson called his Secretary of Commerce Luther Hodges and said, "Lady Bird wants to know what you're going to do about all those junkyards along the highways." Throughout the president's push for beautification, he was clear in associating his wife's interests with the impetus for producing a law, exclaiming that he wanted one passed "for Lady Bird" and telling his cabinet, "You know I love that woman and she wants that Highway Beautification Act," and, "By God, we're going to get it for her." In his state of the Union address on 4 January 1965, he told the nation that a "new and substantial effort must be made to landscape highways to provide places of relaxation and recreation wherever our roads run."[61]

In an address to Congress one month later, the president explained the rationale for beautification. In reference to junkyards, he exclaimed that the nation had to eliminate or screen "unsightly, beauty-destroying junkyards and auto graveyards along our highways. To this end, I will . . . recommend necessary legislation to achieve effective control, including Federal assistance in appropriate cases where necessary."[62] The president asked Congress to enact a bill that would give the federal government power to require states to ban the "beauty-destroying" billboards and junkyards. States failing to do so would lose their federal highway funds. After much debate throughout the summer and fall of 1965, Johnson signed the Highway Beautification Act on 22 October 1965, exclaiming it would "bring the wonders of nature back into our daily lives."[63]

The debates over the Highway Beautification Act of 1965 produced a forceful lobbying effort by ISIS opposing the measure. President Johnson's recommendations troubled scrap dealers. For the first time, the federal government threatened to expand its regulation of scrap firms to include not simply aspects of trade but also the ways that the industry processed and stored its materials. Moshe Oberman, editor of *Scrap Age*, objected to the proposed bill, asking "what is a processor supposed to do when a new highway passes close to his yard and at a higher level? Put a roof on it? In some areas of the country it would be necessary to grow giant sequoias to screen it effectively with living growth. No fence could be as high as required."[64]

Congress debated highway beautification through the summer of 1965. The scrap trade associations began a concerted lobbying campaign to minimize the effects legislation would have on their businesses. The campaign used themes the associations had used for decades—that scrap dealers provided a unique and vital service as agents of conservation and that this activity, unsightly as it may be, had to be performed.

M. J. Mighdoll, executive vice president of the National Association of Secondary Material Industries, Inc. (NASMI), chafed at the use of the term "junk" in describing the industry's products. Testifying before the Public Works Committee of the Senate in hearings on the Highway Beautification bill, he stated "the time has come for our nation's leaders to clarify the terminology relating to an industry that this country found it could not do without during years of war and which it now finds essential to its industry capacity during years of peace." Referring to the industry's products as "scrap" and "secondary" materials, he noted the total output of metals, paper, textiles, rubber, and plastic scrap processors made up a $5 billion annual volume. This hardly is "junk" he stated, particularly to the many industries dependent on these secondary sources for their raw material supply.[65]

The use of the term "junk" in the legislation emphasized negative connotations of the term and induced scrap dealers to consider how they identified themselves. Concern with the industry's reputation led to a new push among industry leaders to urge firms to stop using the word "junk" in describing their work. A front-page editorial in *Scrap Age* claimed that the industry should realize that the word

"junk" was unsuitable for its activities. "The nearly twenty state leg-
islatures which considered measures to control the unsightliness of
'junk' yards—and now the US House of Representatives considering
a similar bill—made no distinction between 'junk' yards and 'scrap
processors.'"[66]

The distinction between "junk" and "scrap" reflected ISIS founder
Benjamin Schwartz's contrast three decades earlier between modern
scrap men who provide a service to the economy and exalted junk
collectors. *Scrap Age* employed Schwartz's rhetoric that "scrap" con-
noted a utility "junk" did not and exhorted its readers to identify themselves as scrap firms, for those who kept the word "junk" in their
corporate names were shortsighted, outdated, incorrect, and harmful.
"The name of this game is 'scrap.' Anything else is a disservice to your
firm, your industry, and your future."[67]

Concern over the industry's image was not limited to questions
of terminology. William S. Story, executive vice president of ISIS, told
his constituents that many needed to voluntarily change their behavior to avoid federal regulation. Recognizing that scrap firms' work was
perceived as a nuisance to many, Story said "if the industry does not
want to be tagged with the junk label, then it must take the necessary
steps where needed to improve its standing in the eyes of the public."[68]

While the trade associations implored their members to limit the
impact of potential nuisances, they lobbied Congress to limit the impact the beautification bill would have on their constituents. NASMI
and ISIS asked Congress to grant compensation to any scrap processors who needed to screen their yards or be removed from their locations. ISIS also successfully lobbied Congress to provide that scrap
processing yards located in industrial areas which could not be effectively screened need not be removed.[69]

The scrap trade associations joined the Outdoor Advertising Association of America (OAAA, the major billboard trade association)
in lobbying Congress. The resulting legislation was much weaker than
what the president and first lady desired, providing "just compensation" for billboard owners whose advertisements were removed and
failing to penalize states for not complying with the law's statutes. At
the bill's signing, President Johnson lamented that the bill "does not
represent everything that we wanted. It does not represent what we
need. It does not represent what the national interest requires. But it
is a first step, and there will be other steps. For though we must crawl
before we walk, we are going to walk."[70]

Historians view the Highway Beautification Act as a minor and ineffective footnote among the Johnson Administration's environmental achievements. Lewis Gould claims that the billboard industry managed to water down regulations of their industry to the point where the law was neither strong nor effective. Subsequent attempts to strengthen the act in the late 1960s proved futile, and environmental groups subsequently saw it as protecting the very business interests it was meant to regulate. Johnson aide Joseph Califano regarded the act as one of Johnson's rare legislative gaffes, one in which Johnson did not take the measure of the opposition to the bill in Congress into account, resulting in what Califano viewed as a weak law.[71]

The Highway Beautification Act is significant in the history of the American scrap industry, however, for two reasons. Unlike its relatively minor impact on the nation's billboards, the law had an effect on junkyards lining the highways. The Department of Transportation estimated that over 3,300 illegal junkyards were removed or screened from view by 1979.[72] The act was also significant in that it represented the first federal attempt to regulate the aesthetic qualities of the scrap industry. Prior to 1965, federal regulation concerned itself with import and export issues, price controls, and transportation. Aesthetics were a municipal or state concern. President Johnson targeted junkyards as a nuisance, and the scrap industry responded with a lobbying campaign based in part upon the rhetoric it used to fight economic regulations. The industry also claimed to work at the forefront of beautification. The Luria-Ford partnership's insistence that their joint venture would reduce eyesores was an attempt to convey to the public and the government that they were on board with the president's attempts to preserve beauty alongside America's highways.[73]

The Throwaway Society

After the scrap drives of World War II ended, Americans once again considered waste handling the domain of a few businesses best kept out of sight and out of mind. That aspect of American waste patterns represented continuity with the prewar years. Yet American waste disposal, including that of consumers, industry, and the state, increased after the war, and the volume and variety of materials discarded grew beyond levels ever seen in the United States. Packaging, food, energy, and commodities ranging from toys to dishwashers to automobiles were consumed more rapidly by a greater number of people than they had ever been. The amount of stuff Americans owned and consumed was far greater than before; two decades of general prosperity fueled by government mortgage and education programs and the expansion of corporations provided more disposable income than the nation had ever seen.

The word "disposable" was apt, as Americans used their income to increase their habit of disposing materials. The automobile industry was perhaps at its most iconic, with distinctive new body types prompting middle-class drivers to discard their old cars after a year or two for the latest sleek, chrome-plated model. Drivers in most parts of the country could take their cars to a McDonald's or White Castle to pick up a paper sack filled with food and disposable cups filled with soft drinks or milkshakes. The amount of packaging materials that went into the average American meal in 1965 was far greater than twenty years earlier. Supermarkets sold more prepackaged foods, from breakfast cereals in cardboard boxes to soft drinks in glass, aluminum, and plastic containers.

What happened to all of this stuff once the consumer had finished using it? Notions of cleanliness had not declined; if anything, the quest for sanitary perfection had become even more ambitious. Modern disinfectants, vacuum cleaners, garbage disposals, and other technologies raised expectations of appropriate cleanliness in the American home. Old, dirty materials were as unwelcome in that home as they had been in 1920, and the old dirty materials of the 1965 home were much greater in size and variety, producing unprecedented solid waste streams.[74]

American cities devised more elaborate systems for handling the wastes. The first sanitary landfill, a lined dump that ostensibly protected the surrounding environment from contamination by the trash, opened in Fresno, California, in 1940. Landfills dotted the postwar landscape, becoming the primary end of the lifecycle of American goods.[75]

Private scrap yards endured, of course, and there the bulk of the nation's old metal lay, either to be reused or to rust. In 1965, the scrap industry was composed of several thousand firms serving a few large industrial customers. Many scrap firms were still owned by the heirs of immigrant junkmen, though a growing percentage followed the lead of Luria Brothers in selling their interests to corporations. Many of the dynamics affecting the scrap industry in the 1950s and early 1960s endured over the next fifteen years; in particular, the trend toward corporate ownership since the end of World War II continued to expand, and the ethic of disposal and general shunning of used materials by society at large defined material usage.

The scrap trade of the postwar era had been affected by the rise in big business during the war; Luria Brothers' rise and eventual prosecution represented the new, complex, and expensive dimensions of material reclamation after 1945. Luria's position as the dominant scrap firm was unique, as was its prosecution for antitrust violations. The firm's transition to corporate ownership was an early signal of the direction of the industry. Corporate ownership of scrap iron and steel firms increased substantially between 1948 and 1963, from 22 percent to 35 percent of all firms.[76] Smaller yards faced competition from large competitors with advantages in customer bases and operating equipment.[77]

The scrap industry, facing an economic landscape of increased costs, had entered by the mid-1960s a period of corporate ownership dependant on both capital from investors and the networks of activity established by the industry's founders. New challenges after 1965, including new federal regulations subsequent to the FTC and highway beautification campaigns, further catalyzed the shift away from the traditional immigrant scrap man to the new reality of the corporate scrap firm.

6 It's Not Easy Being Green

To say that Michael Bloomberg entered his term as mayor of New York City at a difficult time would be an understatement. Elected shortly after the attacks on the World Trade Center, Bloomberg assumed his duties in the wake of tragedy and severe budget shortfalls. He attempted to find savings wherever he could in his first few weeks in office. Bloomberg's administration announced in April 2002 that one of the cutbacks in the city's budget would be the elimination of curbside pickups of glass and plastic containers. New York City Sanitation Commissioner John J. Doherty said that while the city would continue to collect paper and metal, other materials were not cost effective. "Unfortunately, the commercial markets for plastic and glass are practically non-existent and most of these items end up in landfills. Collection and disposal of glass and plastic costs the City about $40 million a year. Given our current budget concerns, it would not be wise to continue squandering money until an economically sound system for glass and plastic recycling can be developed."[1]

The outcry from environmental activists, local officials, and residents was swift and loud. "To stop recycling would be to turn the clock backward," said Suzanne Shepard of the New York chapter of the Sierra Club. "Recycling and waste reduction are the cornerstones to reducing this city's waste stream."[2]

City council member Michael McMahon, a Staten Island Democrat and head of the council committee on sanitation and solid waste, expressed that he was very discouraged that the administration used the budget problems as "an excuse" to kill a program that was helping to reduce the city's garbage problem. "I'm very discouraged that in this tough budgetary time, they use that as an excuse to kill the program."[3]

Bronx borough president Adolfo Carrion Jr. said, "I think people are sort of in shock." He remarked that the program was popular and important. "Here we've been doing this public education campaign, talking about conserving water, reducing waste. We told people this is for the greater good, we passed laws requiring citizens to participate in protecting the environment, then suddenly we say we're going to give up because it doesn't quite balance the ledger."[4]

Much had changed in the ways in which Americans viewed material reuse between 1965 and Bloomberg's 2002 announcement. No urban politician would decry the suspension of a city's recycling pickup programs in 1965, for there were no such things then. Public collection of old materials for anything other than disposal ceased with the end of World War II, yet such programs were common in 2002 and seen as valuable in protecting the environment. This too was new; scrap reclamation had a history of being perceived as being good economic sense, but whatever most people thought of scrap and environmental conditions was negative. Scrap yards befouled the environment; collections of scrap had to be walled off from public view according to the provisions of the 1965 Highway Beautification Act. What had changed that made New Yorkers consider public curbside pickups of recycled glass, plastic, metal, and paper important to their environment? How had recycling, which in 1965 was the exclusive domain of private firms, become a public service? How did the new public concern with material reuse change existing salvage operations?

A New Ethic

Though the growing perception of recycling as a benefit to the environment after 1965 was a break with past patterns, much about American reuse of materials represented continuities with the immediate postwar era. Many environmental advocates at the end of the century prized the return of postconsumer materials to industrial production as a means of reducing pollution. This value was not

Figure 6. Recycling bins, University City, Missouri, 2004. Millions of Americans now sort recyclable papers, cans, and bottles for pickup in the hopes that this behavior will reduce the impact of garbage disposal. *Photograph by Jason Baldwin.*

universal; indeed, it generated resistance from producing industries and criticism from journalists skeptical of the ecological benefits of keeping recyclables out of landfills. The value was widespread enough, however, to alter the waste disposal practices of millions of Americans by the end of the century.

Recycling, a term introduced by the petroleum industry in the 1920s to conserve its own wastes, gained new significance as state and local governments tried to limit the wasteful effects of mass consumption through curbside recycling programs. Under these programs, trucks passing through residential neighborhoods picked up bags or boxes of sorted postconsumer materials, primarily newsprint and packaging such as cans, bottles, and wraps.

The cultural value of recycling translated into political action at the state and municipal levels in the 1970s and 1980s. In 1972 Oregon became the first state in the nation to pass a bottle bill designed to increase the recycling rate and decrease litter by giving consumers a financial incentive to recycle. California, Connecticut, Delaware, Iowa, Maine, Massachusetts, Michigan, New York, and Vermont adopted similar measures between 1973 and 1987.[5]

Several municipalities established curbside recycling programs in the 1970s and 1980s, initially picking up newspapers. The city of Seattle initiated one of the earliest curbside recycling programs in 1969; the nonprofit Berkeley Ecology Center, also founded in 1969, started the Bay Area's first curbside recycling program four years later. In the urban Northeast and along the West Coast, such programs were ubiquitous to the point that municipalities without curbside recycling programs were characterized as the exception rather than the rule. The 1980s saw recycling enter the mainstream in American urban life. At the start of the decade, fewer than 140 communities across the nation had established curbside pickup services; by the 1990s, well over a thousand communities had implemented such programs. In 1989, over 10,000 recycling drop-off and buy-back centers existed in the United States, and more than 7,000 scrap processors were in operation.[6] Between 1970 and 1990, recovery of municipal solid waste as a percentage of total municipal solid waste generated jumped from 6.6 percent to 16.2 percent. Much of the material recovered consisted of glass, metal, paper, and plastic bottles, cans, and other packaging materials found in many consumer products.[7]

In addition to the laws promoting recycling of beverage containers and newspapers, by the early 1990s ten states enacted depositlike requirements for auto batteries. These laws required that consumers buying replacement auto batteries return used batteries or pay deposits of between five and ten dollars. The battery laws required wholesalers to accept returned batteries from retailers, establishing regulations on the market to induce recycling.[8]

The bottle bills spurred an increase in recycling, and over the next three decades a series of laws and regulations led to increasing levels of recycled paper and metals. The federal government's regulations on recycling its own office paper, in fact, led to substantially elevated recycle rates for paper. In addition, companies such as Second Nature and Patagonia began marketing paper products, glue, and clothing explicitly made out of recycled materials that concerned consumers could purchase and feel they were reducing burdens on the environment. The change in behavior implicit in the recycled-goods market was selecting particular brands to buy rather than changing the ways in which people used materials once they purchased them. In this way, the society of conspicuous consumption had refined itself to consume conspicuously, yet in a more responsible manner.

As recycling gained widespread acceptance as an environmental good, several nonprofit and public entities became involved as collectors and sorters of many recyclables. Many cities issued residents separate cans or bins for recyclable cans, bottles, and papers. Pittsburgh's supermarkets gave customers blue plastic grocery bags that could be reused in the city's curbside pickups.

As in World War II, material sorting and reuse became an accepted part of mainstream American society. Unlike World War II, the reasons for the recent acceptance of recycling were not temporary; packaging industries were not in immediate need of new materials, as war production industries had been. Curbside recycling grew in acceptance because recycling was now perceived to have environmental benefits. Recycling kept materials out of landfills and incinerators, reducing human impact on the land and air. A responsible environmentalist was now expected to sort through one's disposables and separate out those materials deemed reusable.

The paradox of American consumption and environmental attitudes in the final third of the twentieth century in part produced the most complicated period in the history of recycling since the immigrant entrepreneurs established an industry at the turn of the last century. Though scrap collecting and processing could now be seen as not only an economic virtue but also an environmental one, scrap firms faced new tensions that built upon existing ones. These tensions may not be understood without first exploring how American attitudes concerning waste did and did not change after 1965.

Rising consumption throughout the 1950s and early 1960s produced a sustained growth in the standard of living of millions of Americans. Consumption also produced unprecedented disposal of materials as garbage and scrap. Sanitary landfills to store refuse away from the public, to the point that people were banned from intruding or scavenging in them, became widely adopted methods of managing refuse after World War II. After two decades of operation, many of these sites, such as New York's Fresh Kills landfill, became immense, visible from several miles away. The inability to escape from the wastes Americans generated produced a movement to reduce the disposal of garbage by extending the life of consumer goods. This concern produced renewed attention on recycling as a public virtue, and, unlike the war effort, this effort was sustained over a long period of time.

Material reuse had become part of the nation's cultural fabric. Consumer recycling continued to gain popularity after 1970, and by the mid-1990s, most American municipalities had some sort of collection program. The heightened popularity of recycling, combined with new municipal, state, and federal regulations designed to encourage recycling, produced a backlash claiming the practice was burdensome and ineffective. Paper and plastic packaging industry advocates, such as the Cato Institute, argued the new recycling laws were costly, ineffective, and unnecessary, most famously in John Tierney's 1996 cover story in the *New York Times Magazine*, "Recycling Is Garbage." Tierney argued that the energy and costs associated with processing and transporting recycled materials made the practice both ecologically and economically wasteful, perhaps "the most wasteful activity in modern America: a waste of time and money, a waste of human and natural resources," and that it would be more affordable and ultimately cleaner to the environment to pitch postconsumer materials into sanitary landfills.[9]

Tierney's conclusions were belied by the presence of a private industry that had long made profits recycling postconsumer materials, but they reflected the opposition to recycling laws by the nation's packaging manufacturers who did not wish to bear any regulatory burdens or costs associated with recycling. Nor did the packing manufacturers wish to bear any costs associated with the end stage of their products. Despite the complications inherent in recycling plastics, the combination of private markets and public regulation expanded the recycling of packaging materials over the last half of the twentieth century.[10]

Advocates and detractors of public collection programs could debate the merits of recycling. What was not debatable was the fact that the presence of legislation to encourage recycling over the final third of the century signaled new attention to recycling as an activity perceived as having environmental, rather than economic, benefits. Recycling, in the words of Samuel P. Hays, evolved into a new ethic that pervaded many sectors of society. This ethic spurred passage of the recycling laws, laws that were favorable to the scrap industry in that they promoted the same activity that scrap firms had engaged in since their inception.[11]

At the same time recycling became an accepted part of solid waste management, a combination of increasingly complex manufactured goods and new concerns over environmental health and toxicity made the work of storing and processing recycled materials more difficult than before the late 1960s. As green activity became important, recyclers old and new found that being green was not easy.

This new recycling ethic was not a simple return to the nineteenth-century stewardship of objects. Public recycling programs prompted the consumer to separate out the materials leaving the household into different streams, but those materials still left the household. These programs did not spur customers to reuse old materials within the household but rather to make sure the materials would not enter the waste stream once they were evacuated. The new programs represented a major change from the throwaway ethos of the 1950s, but they remained part of the impulse to keep the household clean by removing old materials.

This is not to say real concerns over American consumption did not lead to efforts to reduce its effects. Grassroots efforts to reuse old materials within the household became more prevalent, including composting, quilting, and recycling objects as art. Composting keeps old edibles in the yard, something that might have been met with revulsion a few decades earlier, when chemical fertilizers were seen as the appropriate way to maintain one's lawn and garden. One may, without much difficulty, purchase an address book or purse made out of old license plates. But these efforts are not as widely practiced as curbside recycling. Americans' tendency to consume materials, energy, and water continue to rank among the highest of any society, even though care is taken to sort aluminum cans, plastic bottles, and news-papers from garbage. Public recycling is environmentalism based upon traditional consumer behaviors.

The Original Recyclers

Consumer recycling programs benefited the scrap industry in two ways. First, most public recycling programs involved contracts with scrap dealers to process and sell the collected materi-als back to manufacturing industries. Second, these programs brought the use of the term "recycling" into the lexicon of the environmental movement. The new municipal recycling programs grew from concerns over the effects of mounting solid waste rather than the business-oriented concern over extending the available resources for production.

The new environmental emphasis on recycling allowed the scrap industry to recast its image as an environmentally friendly industry. Scrap firms had marketed their activities as being crucial to the con-servation of vital national resources, thus providing an important eco-nomic service. In the 1970s and 1980s, the trade associations embraced the rhetoric of environmentalism, asserting that their role reintroduc-ing materials into industrial production made them original recyclers who protected the environment by minimizing solid waste disposal. The National Association of Secondary Material Industries renamed itself the National Association of Recycling Industries (NARI). The as-sociation, originally named the National Association of Waste Mate-rial Dealers in 1913, had twice shifted its identity in order to make appropriate rhetorical claims for the times, evolving from waste ma-terial dealers to secondary material dealers to recyclers. ISIS retained its name throughout the 1970s, though it too adopted the rhetoric of recycling, calling its members "Original Recyclers" and casting their efforts as vital to the environmental movement. The Institute also at-tempted to co-opt the language of the highway beautification move-ment. ISIS called its members' efforts ones of "conservation, reclamation and beautification" in a pamphlet published in 1974.[12] When ISIS and NARI merged in 1987, the new organization called it-self the Institute of Scrap Recycling Industries (ISRI), merging the struc-ture of the established Institute with the resonance the word "recycling" now had in American politics and society.

The junked automobile, the pop can, the bottle, and the old news-paper were by 1970 symbols of conspicuous consumption that envi-ronmental groups in towns across the United States attempted to keep out of growing landfills through recycling campaigns. Plastic and card-board packaging proliferated, as did its disposal in landfills and un-sightliness as litter. Between 1960 and 1980, the amount of paper and paperboard packaging material—including corrugated boxes, milk car-tons, folding cartons, bags, sacks, wrapping paper, and other paper packaging—disposed of in municipal solid waste rose from 14,124,000 tons to 26,314,000 tons.[13] Regulation of automobile graveyards, as dis-cussed in the previous chapter, marked the beginning of federal regu-lation concerned with the aesthetic impact of waste material handling. The automobile and beverage containers became the first mass con-sumer items in peacetime to have specific municipal and state legis-lation enacted to encourage their reclamation.

How did these changes affect existing salvage efforts? Scrap firms had undergone significant changes since World War II, and many of the family-owned businesses had consolidated into corporate firms that were relative giants. After 1965, consolidation continued. Other changes included a spread of firms into the growing industrial centers of the Sun Belt, particularly Texas and California.

Iron and steel remained the most widely traded secondary mate-rials in the late twentieth century; the ferrous scrap industry grew in terms of both number of individuals employed and volume of scrap sold. A study by the Battelle Institute estimated the total employment of the ferrous scrap processing industry at forty thousand workers na-tionwide in 1972.[14] The opportunity structure of the industry evolved to the point that scrap firms operating in the 1970s had tenuous links to the immigrant scrap firms of the early twentieth century. Most scrap processing firms remained small operations employing fewer than ten people, with a minimal amount of processing equipment and shipping less than one thousand tons per month. Their share of the volume of scrap traded, however, shrank as large broker/processors owned by corporations grew larger.[15]

The domestic market for scrap iron and steel maintained its spa-tial focus over the course of this period. Scrap firms operated in ev-ery state of the union in 1970, yet the nexus of activity remained situated in Pennsylvania, New York, and New Jersey. Firms located in those three states accounted for 30 percent of the nation's total scrap trading activity. The industrial midwestern states of Ohio, Michigan, Indiana, Illinois, and Wisconsin accounted for 26 percent of volume traded, and California, Oregon, and Washington accounted for 14 per-cent of the nation's ferrous scrap trade.[16]

The export trade, a source of controversy at the beginning of World War II, rose throughout the 1960s. By 1970, one-quarter of total purchased ferrous scrap in the United States went abroad.[17] Japan accounted for the most sales of exported scrap, just as it did in the 1930s. The industrialized nation with few domestic resources of its own purchased 49 percent of all American ferrous scrap exports in 1970. Spain was the second largest foreign consumer, accounting for 14 percent of the foreign market. Mexico, which was attractive to steel manufacturers for its proximity to the United States and low labor costs, accounted for 8 percent.[18]

At the same time ferrous metal enjoyed rising demand abroad, it retained its status as America's most widely traded scrap commodity. Domestic sales of ferrous scrap totaled 35,804,000 tons in 1965.[19] The customers for ferrous scrap firms continued to be large industrial producers such as mills, foundries, and automobile manufacturers, distributed throughout the nation, but particularly in the traditional industrial areas of the Northeast and Midwest. The distribution of customers in California was a result of continued industrial expansion in that state since World War II.[20]

Automobiles remained popular products to use and discard, as evident in the growing number of junked motor vehicles accumulating in scrap yards. The annual number of scrapped cars approached eight million in 1970, almost double the figure for 1955 and more than two million more cars scrapped annually than were scrapped in 1960.[21] The abundance of discarded vehicles provided an impetus for legislation to abate the aesthetic impact of junked cars in plain view from highways, but it also provided a supply of salvageable material. The proliferation of automobile graveyards alongside the nation's roadways represented a growing ethic of consumption that included refrigerators, air conditioners, stoves, dishwashers, and other items of mass consumption that by their disposal provided potential materials to trade.

The largest scrap firm in America during the postwar period entered the 1970s diminished but still gigantic. Ogden Corporation lost its Supreme Court appeal of its subsidiary Luria Brothers' antitrust sanctions in 1968. The company remained a major broker and exporter of iron and steel scrap, with offices and yards in several cities. Its relative dominance over other broker/processors diminished from its early 1960s peak as large brokers at the beginning of the 1970s included Commercial Metals Company, Schiavone-Bonomo, the David J. Joseph

Company, Proler Steel Corporation, Roblin Industries, and Steelmet, Incorporated. ISIS estimated a total of approximately 150 such broker/processors operated nationwide in the early 1970s, and virtually all of the ferrous scrap traded on the open market passed through their yards.[22]

Several of these firms, including Commercial Metals and Proler Steel, were steel producers that had moved into the scrap processing and selling business by purchasing existing scrap firms and absorbing their expertise and contacts. Acquisition of scrap firms was one way producing companies exerted greater control over purchase and sale of their scrap; relationships such as the one in which the Ford Motor Company allowed Ogden to build and operate scrap sorting and processing operations on the sites of Ford's own plants was another. Both kinds of relationships blurred the line between producing industries and the scrap dealers that served them; one option consisted of clear vertical integration, while the other was a more complex, interdependent relationship between dealer and customer.

The scrap industry's centers conformed roughly to the size of American metropolitan areas in 1980. The three largest centers of scrap sales and firm location were the nation's three largest metropolitan areas, albeit not in exact order. New York remained the area hosting the most firms, with Chicago edging out the Los Angeles metropolitan area for second place, even though Los Angeles was now the nation's second largest city. Chicago enjoyed the largest sales of any metropolitan area, followed by Los Angeles, where the market included steel works in Fontana and shipping and defense industries throughout the region. New York was third. The geographic spread of the scrap industries was evident by the growing number of firms in Texas and California. The San Francisco Bay Area was ninth in number of firms and seventh in sales receipts. Dallas–Fort Worth was seventh in number of firms, and Houston was eighth in number of firms and ninth in sales.

Traditional centers of the scrap trade, including Philadelphia, Detroit, Pittsburgh, and Cleveland, lost ground between 1960 and 1980 as deindustrialization reduced demand from local customers. Scrap firms, dependent on income from producing industries, operated where there were customers. Exports out of New York, Los Angeles, and Houston also affected the growth of those markets. Many exports to Japan sailed out of Los Angeles and Long Beach. Houston's growth during this period was in part due to the growth of the Mexican steel industry, as well as labor costs lower than those in the industrial Midwest.[23]

The large brokers developed operations resembling Luria's, with offices and yards spread throughout the country, hundreds of employees, and sales of over $100 million of business annually.[24] These firms had the capital to invest in equipment that allowed them to process high volumes of scrap. Shredding and fragmentizing equipment, capable of producing uniform scrap from automobiles and other complex consumer goods, became commonplace in large processors' yards.[25]

Scrap Yards to Minimills

As large corporations continued to dominate the steel industry in the 1950s and 1960s, technological changes that would undermine that dominance were already under way. European producers began using the electric arc furnace and continuous casting steel production in the 1950s. The electric arc furnace and continuous-casting process developed after the large integrated mills were built. The old mills were invested in old technology and were reluctant to modernize, but new, small firms had no such sunk costs and were free to adopt new techniques. Furthermore, the new processes were adaptable to small-scale production.

The new steel production techniques of the 1950s and 1960s that allowed mills to reuse factory scrap gave new steel fabricators economic advantages over the industry giants, who declined in the 1970s and 1980s as small scrap firms opened minimills. Continuous-casting and electric-furnace steelmaking provided the technological basis to permit minimills to compete against large, integrated companies. The minimill, a type of steel production that required less real estate or manpower than large mills did, and which could accept charges composed entirely of scrap iron and steel of various qualities, gave scrap yard owners a viable option to enter the steel production industry. Just as Jim Levinson's children had transformed his scrap yard into Levinson Steel in the late 1920s, scores of yard owners shifted into production in the 1970s and 1980s.

The large integrated mills that employed thousands were supplanted by small producers in the United States, as well as Mexico and overseas. The steel industry began to move away from large, integrated systems. This led to a decline in the major steel centers of the United States; Pittsburgh experienced sharp reductions in its steel production between 1970 and 1985; by the end of the 1980s, steel production had virtually disappeared from the metropolitan area, replaced as the dominant employer by services in health care and education. Gary, a town built by U.S. Steel and named for its president, did not see steel production eliminated, but its productivity and employment were reduced by more than one-third by 1990.[26]

The rise of the minimill provided both challenges and opportunities for the ferrous scrap industry. The electric-arc furnace initially had a negative effect on the scrap market. Although it allowed for virtually all ferrous scrap to be recycled, the effect was that mills no longer had excess scrap—they could use everything they had in production. This led to a reduction in demand for outside sources. The restructuring of the industry as initially overseas steel adopted the more efficient processes and American production caught up and surpassed consumer demand in the late 1950s led to prolonged slumps in the domestic ferrous scrap market.

Many scrap dealers recognized opportunity in the advent of the minimill. With supplies of usable scrap on hand and contacts to get more, a successful dealer could invest in a new, modern minimill and shift from being a supplier to being a manufacturer. The shift was an example of a flexible business structure and a continuation of a move out of the scrap industry that was advantageous for both economic and cultural reasons. Economically, smaller firms found it hard to compete with large brokers who could process large volumes of material and who had established relationships with giant industrial customers, relationships that could not be easily displaced. Culturally, the stigma relating to scrap and waste handling endured, complicated by the new dimensions of recognized environmental quality and subsequent federal regulations of environmental aspects of the scrap industry. Given this context, many scrap firms became minimills, continuing to recycle steel but engaging in the production of new materials rather than collecting "wastes" to convert.

Minimills in the United States and throughout the world proliferated; the *Wall Street Journal* estimated sixty minimills were in operation in the United States in 1981. The largest ones were able to produce 1.5 million tons of steel each year. Boasting more efficient production techniques—the electric arc furnace and continuous-casting facilities required about 9.9 million British thermal units (BTUs) of energy to produce a ton of steel at a time when integrated steel facilities used about thirty-five million BTUs per ton—and lacking the contentious and expensive labor relations of the large producers, minimills flourished. At a time when large steel companies were on the decline, minimills accounted for as much as 15 percent of domestic steel output.[27]

Even scrap yards that did not convert their operations to small-scale steel production benefited from the proliferation of the new mills. Minimills typically relied on local sources of scrap, consuming material that might otherwise lay dormant in scrap and automobile yards as large mills' demands for purchased scrap declined.[28]

New Materials, New Markets, New Problems

The development of minimills was one example of how technological changes provided scrap firms with new opportunities. Other technological changes complicated methods of doing business. The changing nature of materials used in industrial production affected what the scrap industry collected and how it processed materials. Composite materials such as plastic on metal, combinations of dissimilar metals, polymers on fibers, and polymeric fiber combinations were materials engineered for various purposes in the 1960s and 1970s. New metal alloys designed for motors, engines, and industrial equipment became commonplace.

The impact of new alloys and combinations of material was twofold. Metallurgists and chemical engineers required secondary materials of high quality to fabricate sound alloys. Thus, demand for high-quality scrap increased. Processors employed new techniques and technologies to sort and process high quality scrap. Fragmentizers and electromechanical machines developed in the late 1960s allowed processors to separate relatively pure metals out of complex postconsumer products.[29] The shredder, designed to process automobiles, refrigerators, washing machines, and other complex objects, gained popularity; the *New York Times* reported in 1972 that the number of shredders in use in the United States had doubled since 1968. One hundred and eight shredders processed 2.6 million automobiles in 1972, quickly separating iron and steel from plastics, fibers, and other residues, producing high-grade ferrous scrap.[30] Processors who could afford to invest in new technologies were able to satisfy customer demand for high quality scrap as the materials in the industrial life cycle became more complex, but such adaptation required substantial investment costs in machinery.

Though ferrous metal continued to have the largest markets in the secondary materials industry, other materials expanded their markets in the 1960s and 1970s. Battelle's estimates of recycling rates for materials in 1969 ranged from a low of 14 percent for zinc to a high of 88 percent for stainless steel. Materials with higher unit prices had higher recycle rates, explaining the 75 percent recycle rate for precious metals. Scrap copper remained a valued commodity; about half of all copper and brass used in American manufacturing in the early 1970s was scrap. Scrap lead was used in similar ratios; secondary aluminum represented about a third of the nation's aluminum used. Secondary gold and paper stock made up about a quarter of all such raw materials in the nation.[31]

Political and economic factors affected recycling rates for various materials. The trade-in policy for automobile batteries boosted the recycle rate for lead, with 45 percent of disposed lead recovered in 1969. The decreased demand for cylinder paperboard in turn depressed the recycle rate for paper (24 percent).[32] A downturn in paper recycling was part of a long trend. Between 1945 and 1970, the amount of paper recycled declined by almost half, from 35,000 tons to 18,000 tons. The decline came despite an increase in paper consumption; waste paper disposal became one of the nation's most significant solid waste problems in the 1970s.[33]

The particulars of industrial manufacturing also affected recycling rates for specific materials. The sacrificial corrosion of zinc from galvanized steel depressed zinc's recycle rate (14 percent). Conversely, brass and aluminum manufacturers adopted continuous casting, allowing them to use charges composed almost entirely of scrap metal. The process required modifications in temperature control, as steel freezes at a much lower temperature than brass or aluminum, but the process was readily adaptable to the industry, bolstering the markets for secondary brass and aluminum.[34]

It would be too simple to sum up the difficulties of recycling in the late twentieth century by quoting *The Graduate's* famous line, "one word—plastics," but new synthetic materials made reclaiming postconsumer and postindustrial materials more complicated. Plastics, including polychlorinated biphenyls (PCBs), joined heavy metals as potentially hazardous materials found in many products scrap recyclers had to process to return metals and paper to the industrial life cycle.

Plastics provided many benefits. Synthetic materials grew in popularity between 1945 and 1970 as they provided light durable and cheap alternatives to rubber, glass, and paper as packaging, components for automobiles and other complex machinery, and tires. They supplanted glass and metals in light and durable beverage containers. The proliferation of plastics reduced demand for scrap rubber after World War II; secondary rubber and paper dealers had difficulty selling their stocks in the 1950s and 1960s, and many left the industry.

But plastics did not provide immediate alternatives as salable secondary materials. The complexities of plastic manufacture made recycling of the materials difficult, and plastic packaging was not biodegradable. Furthermore, some plastics were carcinogenic, and the various grades of plastic in wide use meant that accidental mixing of plastics could easily ruin a batch of recyclables. Plastics represented less than 2 percent of the United States' collectable waste in 1970, but as they grew in popularity in the 1970s and 1980s, difficulties recycling the materials led to demands from activists to reduce their use in packaging.[35] The science that created new materials in the postwar era also created new unintended consequences for human and environmental health.

Against the Earth

Biologists and chemists found evidence that pollution associated with consumption comprised toxic threats to the air, water, animals, plants, and human life. The publication of Rachel Carson's *Silent Spring* in 1962 brought attention to the dangers of synthetic pesticides to not just insects but also humans, plants, animals, and the earth itself. Carson compared the dangers of pesticides to those of the most feared threat of the era—nuclear fallout. The book's impact was tremendous. *Silent Spring* became a best seller, catalyzing attention to the dangers of polluting the environment. Public demands for measures to curb pollution increased over the decade.[36]

This new emphasis on protecting the environment had roots in the past. The conservation ethic that produced efforts to protect forests and lands under Theodore Roosevelt's administration had durable effects on American conceptions of managing nature. Federal agencies such as the Forest Service and advocacy groups such as the Sierra Club and Audubon Society continued in the 1960s to manage (or advocate) federal lands with the rationale of protecting them for future use. Efficient conservation of resources settled into the mainstream of Republican political values and corporate boardrooms, and became rhetoric that the scrap industry eagerly adopted. Conservationism allowed scrap dealers to emphasize their service of providing materials and skills to manufacturing industries. Scrap firms used their rhetorical claim of being "vital agents of conservation" well into the 1960s. To the extent that scrap firms maintained longstanding relationships supplying mills and factories with secondary materials, that claim was successful. Federal regulations prior to 1965 assessed scrap firms on their economic virtues, and so the rhetoric of conservation was sufficient when the industry lobbied Congress on matters of tariffs, interstate commerce rates, and price ceilings.

The body of federal and state environmental law that emerged after 1965 reflected a new set of values, including a respect for the quality of the natural environment from an amenity and aesthetic perspective and a concern for health in relationship to the environment.[37] As public support for reducing pollution grew, demands for action focused on local, state, and federal levels of government. The proliferation of air and water pollution made regulation on local and state levels difficult, and demands for pollution controls emerged on the federal level. This resulted in a series of federal laws in the 1960s and 1970s, including the Clean Air Act of 1970 and the Clean Water Act of 1972. These laws became known as "command and control laws" that attempted to limit the levels and scope of pollution.[38]

Environmentalism, and the regulations inspired by environmental concerns, drew swift opposition from affected industries. Manufacturing industries, including pesticide manufacturers, derided *Silent Spring* as hysterical and lobbied to fight passage and enforcement of command and control legislation, claiming that the dangers of industrial pollution were overemphasized and the costs of regulation were unduly burdensome.[39]

When Lyndon Johnson signed the Highway Beautification Act of 1965, he ushered in a new era of federal policy involving scrap materials' relationship to the nation's environment. The first major federal laws dealing with environmental quality were the water pollution statutes of the late 1940s; those first attempting to moderate air pollution were enacted in the 1950s. The debates over highway beautification in 1965 brought new attention to the enormous quantities of solid wastes generated by the consumer society. Environmental groups and legislators recognized waste as a national problem, and in 1965 Congress passed both the Highway Beautification Act and the Solid Waste Disposal Act. The second act created the Office of Solid Wastes and provided the federal government with a more formal role in regard to municipal wastes. The Solid Waste Disposal Act provided funds for research investigation and demonstration and for technical and financial assistance to state and local governments and interstate agencies in the planning, development and conduct of disposal programs.

The Solid Waste Disposal Act's most important impacts were to stimulate research and to inspire state government activity. In 1965, for instance, there was no state-level solid-waste agency in the country, but by 1970 forty-four states had developed programs. During the 1970s, however, the focus of federal legislation moved from research into conventional methods of solid-waste disposal toward the reuse and recycling of resources, as reflected in the passage of the Resource Conservation Act of 1970. Other federal legislation passed during this period that related to solid waste problems included the National Environmental Policy Act of 1969 and the Environmental Quality Improvement Act of 1970.[40]

The Nixon Administration established the U.S. Environmental Protection Agency (EPA) as the chief federal enforcement organization of the new spate of environmental regulations. Establishing operations on 2 December 1970, the EPA absorbed environmental programs formerly scattered throughout various federal agencies. These included the solid waste programs of the Bureau of Solid Waste Management, Bureau of Water Hygiene, and portions of the Bureau of Radiological Health, all programs coming from Health, Education, and Welfare. The president argued that the reorganization would treat "air pollution, water pollution and solid wastes as different forms of a single problem."[41]

The EPA's mission included protecting the environment against emissions of contaminants through hazardous materials and hazardous wastes. Hazardous wastes were classified as waste materials having some of the following characteristics: ignitability (wastes that pose a fire hazard during routine management); corrosivity (wastes requiring special containers because of their ability to corrode standard materials or requiring segregation from the other wastes because of their ability to dissolve toxic contaminants); reactivity (or explosiveness); toxicity (wastes that, when improperly managed, may release toxicants in sufficient quantities to pose a substantial hazard to human health or the environment).[42]

The complex nature of manufactured goods meant that many items given to scrap dealers possessed materials now defined as hazards. Many consumer products, for instance, including automobiles, refrigerators, and air conditioners, contained potential hazards. Air conditioners used Freon and potentially carcinogenic CFCs. Automobile batteries used lead and other heavy metals, and automobile bodies contained petroleum that could contaminate groundwater.

Section 212 of the 1970 Solid Waste Act required that the EPA undertake a comprehensive investigation of the storage and disposal of hazardous wastes. This led to a report to Congress in 1974 on the disposal of hazardous wastes and eventually, in 1976, to the passage of the Resource Conservation and Recovery Act (RCRA). Congress passed the RCRA to provide assistance to state and local governments to improve solid waste management techniques. Its goals were twofold: to minimize the amount of solid waste generated by maximizing recovery efforts of recyclable materials and to minimize the amount of hazardous waste generated and disposed of by encouraging substitution of processes and materials, recovering hazardous materials and treatment. The act gave the EPA the authority to control hazardous waste from cradle to grave, including the generation, transportation, treatment, storage, and disposal of hazardous waste.[43]

Related to the new environmental regulations were new concerns about safety in the workplace. The Occupational Safety and Health Administration (OSHA), created in the Department of Labor on 28 April 1971, the date the Williams-Steiger Occupational Safety and Health Act (P.L. 91–596) became effective, required employers to "furnish . . . a place of employment which is free from recognized hazards that are causing or are likely to cause death or serious physical harm to employees." Employees could trigger investigations by the agency with complaints of workplace dangers, meaning an industry with historically poor safety records was now subject to federal investigations due to the hazards it presented to both the environment and its own workers.[44]

The dangers of handling scrap increased firms' labor costs over the postwar era as insurance premiums rose. ISIS and NASMI began campaigns to educate their members on workplace safety and ways in which dangers to labor could be minimized. ISIS President Joseph S. Schapiro remarked, "for a scrap processor with approximately fifty men to have a bill of over $200,000 a year for insurance would really make anybody sit up and take notice." Schapiro also warned processors that insurance carriers could not only raise premiums on accident-prone businesses but also cancel their policies. Scrap firms' concerns regarding safety stemmed from both federal regulations and market forces affecting their insurance premiums.[45]

The new federal environmental regulations brought challenges to the way the scrap industry operated. The Battelle Corporation began doing comprehensive surveys of the ferrous scrap industry at ISIS's behest in the 1950s. These surveys were part of the ongoing professionalization project of the trade associate, attempting a rational means of identifying the major concerns and challenges to the industry and advocating ways of solving its problems. By 1972, these concerns and challenges included federal environmental regulations. Potential degradation of the environment shaped the industry's relations with federal regulators. The 1972 Battelle study indicated that the industry was more cognizant of perceptions of environmental safety and aesthetic quality in the early 1970s than it had been previously, concluding "the need for better pollution control is increasingly evident and results are being demanded. Aesthetic values must be considered in laying out processing plants."[46] The study concluded that suburban sprawl and public concerns about environmental quality produced "protests, publicity, legislation, and pressure" that limited the location sites available to scrap processors or forced existing firms to change sites.[47]

Federal environmental legislation regarding recycling culminated in the final year of the Carter administration. Concern over the effects of land disposal of hazardous waste grew in the 1970s. Research into the effects of groundwater contamination by hazardous waste sparked several municipal and state warnings, most famously the evacuation of the Love Canal neighborhood of Niagara Falls, New York, after chemicals from an industrial dump established by the Hooker Chemical Company next to a residential neighborhood began to migrate into yards. Early in 1980, acting under the requirements of RCRA, the EPA announced new regulations implementing cradle-to-grave controls for handling hazardous wastes. On 11 December of that year, Congress passed the Comprehensive Environmental Response, Compensation, and Liability Act (CERCLA), more commonly known as Superfund. This law provided broad federal authority to respond directly to releases or threatened releases of hazardous substances that might endanger public health or the environment, billing the parties responsible for the pollution for the costs of cleaning up the hazards. The fifteen years between 1965 and 1980 saw the federal government establish several laws designed to address the dual problems of solid waste and hazardous waste management.[48]

Responsible Parties

New environmental regulations affected scrap firms' processing techniques and liabilities. Under the provisions of the Resource Conservation and Recovery Act, manufacturers were not held legally accountable for the potentially hazardous wastes of their products. Materials in scrap yards, however, were judged to be at the end of a product's life. Therefore, they were waste materials, even if they were to be reused in industrial production. Thus, the regulatory burdens that come with handling hazardous wastes within scrap materials fell on scrap processors, not manufacturers. These regulations included fines, lawsuits, and criminal penalties.[49] Companies now had to assess how their processing and storage methods affected the land, air, and water, as well as their workers. The trade associations worked with the EPA to investigate how the industry could best minimize the nation's waste of salvageable materials while complying with federal regulations.

Most regulations affecting the industry at the municipal and state level were aimed at local scrap processors and dealers rather than at recycling or the recycling industry as a whole. Municipal regulations concerning scrap yards were similar in scope to the zoning regulations established after World War I in their attempts to separate yards from residential areas. Yards situated in inner cities or on the suburban fringe were often relocated because of new highway construction or changes in zoning. The Battelle study concluded that zoning discriminated against smaller scrap firms, which "often leave the industry and go out of business when faced with the problem of moving to a new location."[50]

Between local zoning laws and federal highway beautification efforts, scrap yards could not hide from the society whose conspicuous consumption produced the need for scrap yards. As with the growing problem of mounting landfills, concern over the aesthetic nuisances produced by scrap yards reflected Americans' discomfort with their own wastes.

In addition to contending with new environmental regulations, scrap dealers continued to face economic regulations, including export embargoes and freight tariffs. Freight rates soared in the late 1960s, causing the trade associations to mount a lobbying campaign to control rates. Freight charges caused many dealers to opt for water transportation, either along the river systems for domestic distribution or across oceans for international trade. River barges, despite being more vulnerable to changes in weather, bypassed the high freight charges of rail transportation. Economist Herschel Cutler joined ISIS in 1968 with the task of convincing the government to establish rates that would not prevent scrap firms from making profits. (He was subsequently appointed executive director of the association.) ISIS's lobbying campaign on freight rates ended with the deregulation of the nation's railroads in the early 1980s, but it had the significant effect of increasing the trade association's budget to lobby the federal government.[51] Other lobbying efforts by the trade associations in the 1970s included a successful effort on behalf of a recycling investment tax credit in the late 1970s and several campaigns regarding workers' compensation and yard safety regulations.[52]

Under Superfund, the EPA directed cleanups of contaminated industrial sites, and the parties held to be polluters of the sites were held liable for the costs of the cleanups. Superfund affected the scrap industry because, although the industry was devoted to recycling, its methods of recycling materials had negative effects on environmental quality. Scrap processors engaged in processes that environmentalists considered unsustainable. The system of reclaiming ferrous scrap from complex consumer products was never a completely closed loop. Aside from materials leaving the cycle through abandonment or export, many materials were disposed of as waste, and scrap processing released wastes into the environment.

Scrap processors separated ferrous scrap from hazardous materials contained in complex consumer products. The automobile is an example of a complex consumer product containing multiple hazards, including PCBs, CFCs, oil, solvents, acids, lead, cadmium, molybdenum, and asbestos. Automobiles with airbags contain the carcinogen sodium azide. Scrapping automobiles released these hazards into the environment. Before the 1970s, processors frequently burned nonferrous materials away from ferrous scrap, contaminating the air. Much of the air pollution arising from iron and steel reprocessing occurs as dust and fume, largely iron oxide. Iron oxide could be recycled if the processor was able to capture it; if not, it was released into the air.[53]

Though burning scrapped automobiles was curtailed in the 1970s (in part because large fragmentizers such as Ogden/Luria's at Ford's Taylor Township facility allowed large processors to take a greater percentage of the scrapped automobile market), scrap processing continued to pose risks to the environment. Cutting and shredding automobile bodies released hazards into the ground and residue from shredded automobile bodies could contaminate the water table. The effects of this activity led many scrap firms into conflicts with federal regulators.[54] The EPA classified many scrap disposal and processing practices as hazardous and dozens of scrap yards were classified as Superfund sites after 1980. Superfund designation led to the contaminated sites closing, followed, often years later, by the EPA removing the toxic materials and billing the sites' owners for the work. The Union Scrap Iron and Metal Company, for example, processed automobile batteries in Minneapolis and Louisiana between 1973 and 1985, spilling acid, lead, and cadmium into the soil. After the company filed for bankruptcy, the EPA directed cleanups at three of the company's sites, costing at least $320,000. The agency attempted to collect the cost of the cleanups from several potentially responsible parties (PRPs), including scrap dealers and automobile graveyards that sold batteries to the Union Scrap Iron and Metal Company.[55]

Many yards designated as Superfund sites were owned by bankrupt firms that left contaminated sites behind. A yard originally owned by the Spokane Metals Company from 1936 to 1983 and subsequently sold twice in seven years to local landowners stored military surplus equipment, automobiles, appliances, and electrical transformers. It closed after an explosive fire in 1987 that released toxic fumes and sickened four firemen. After the fire, the EPA removed 140 drums of hazardous liquids and solids, including PCBs and 140 cubic yards of asbestos. Subsequent investigation revealed elevated levels of liquid mercury, cadmium, and lead in the soil at the site, prompting a cleanup of over 200,000 cubic yards of soil. Seven years after the fire, the EPA designated the yard a Superfund site and began the process of getting the yard's present and previous owners, suppliers, and customers to pay for a cleanup exceeding $2.8 million.[56]

The practice of making yard owners past and present responsible for cleanups, even when their activities had not contributed to the contamination of sites, led the industry to lobby for changes in Superfund. In one case cited by ISRI, eight hundred companies, mostly small scrap firms, were named as PRPs for selling lead-acid batteries to a lead refining firm that polluted a Pennsylvania site and later went bankrupt. Most of the involved firms collected discarded car batteries and sold them, sometimes through other companies, to the Pennsylvania firm. The estimated cost of remediating the site was $35 million, for which all of the suppliers named as PRPs were ostensibly liable.[57]

ISRI argued that the law made scrap dealers PRPs unduly liable for cleanups, even in cases where their scrap was sold to consumers who then contaminated the environment by processing or disposing of it. One scrap dealer held as a PRP in the Pennsylvania case claimed, "we do not have control over our consumers' storage procedures or processing methods, and, therefore, we should not be asked to fund the cleanup of a problem we did not create."[58] After a lobbying effort spanning most of the 1990s, Congress revised the law in 1999 with the Superfund Recycling Equity Act, exempting certain recycling activities from Superfund liability, particularly removing the threat to scrap firms of being named PRPs when their customers disposed of hazardous materials improperly.[59]

Scrap dealers viewed Superfund as the most prominent of several federal regulations hindering small firms. Criminal and civil penalties stemming from the nature of their materials were not new risks for scrap firms. Waste paper and rag dealers had been subject to local regulations to limit the fire and infectious disease hazards posed by their stocks since the early nineteenth century. The new regulations of hazardous wastes, however, made most scrap material handling subject to federal penalties, on the basis that the scrap yards were the final stage of the cycle for hazards contained in scrapped goods. Scrap processors complained that holding them (rather than manufacturers, who put the potentially hazardous materials in products in the first place) liable for hazardous wastes prevented them from doing business and reintroducing materials into industrial loops.[60] Many scrap dealers who retired or sold their businesses cited costs associated with federal environmental and workplace safety regulations as factors in their decisions to leave the industry.

The growing consolidation of the industry under corporate ownership since 1950 prevents attributing the demise of small, family-owned firms solely to Superfund or any of the federal environmental regulations enacted after 1965. Scrap dealers, however, lamented the burdens of environmental regulation. One veteran scrap dealer exclaimed in 1986 that the EPA was a concern "hanging over us like a guillotine." Veteran scrap dealer Fred Berman complained that it was too difficult too keep up with the changing regulations and costs of compliance. "If my dad had to put up with all the conditions today, there'd be no scrap business."[61]

From Conservation to Recycling

As the regulatory structure surrounding the industry became more complex, scrap dealers continued to reshape their identities. "Scrap" firms distanced themselves from "junk" dealers, as they did during the beautification debates, using a strategy long followed by an industry claiming to conserve resources rather than handle wastes. Modifications to that claim were necessary in the 1970s. The industry's attempts to use the rhetoric of conservation, long effective, could not address concerns over hazardous waste disposal. Scrap dealers' attempts to define themselves as agents of vital economic interests did not address new concerns over environmental contamination, thus the public relations strategies used for decades were insufficient tools for lobbying political bodies, customers, and the public. The Battelle study recognized the rhetorical shift, noting that scrap firms functioned "in the traditional *economic* environment," but more recently concluded that economic gain was no longer the sole driving force for recycling and additional considerations included "(1) *improvement of the living environment*, and (2) increased national concern with *conservation of natural resources*."[62]

The scrap industry found that one aspect of their longstanding claim to address the second dimension listed by the Battelle study—conservation of natural resources—could also address the first claim—improvement of the living environment—and speak to the nation's "social gain" as well as its economic gain. Combining social and economic gains made recycling very popular as an alternative to landfills. As hundreds of communities established either drop-off or pick-up programs for recyclables, opportunities to profit from these collections increased. For the most part, however, the scrap dealers who had managed the return of salvaged materials to industry for decades were not the agents contracted to handle municipal recycling collections. Instead, cities contracted with garbage haulers to handle both refuse and recyclables.

Waste management became big business in the 1990s. Several cities privatized their solid waste management systems, entering into contracts with giant private haulers to handle both garbage and recyclables. Private firms had entered into contracts with cities for decades, but where once ten to twelve thousand independent companies handled municipal solid waste services, four national corporations now dominated the market. As in the scrap trade, garbage hauling consolidated as investment costs, technological innovations, and regulatory complexity put small companies at a disadvantage. Garbage giants such as WMX and BFI integrated transportation, landfills, and incinerators. They also opened recycling plants, assuming the contracts of many urban collection programs. In many instances, the scrap firms that had decades of experience returning salvaged materials back to industry were kept out of the cycle of municipal recycling; urban equating of garbage and salvage as solid waste led cities to augment existing solid waste management with additional contracts.[63]

As the garbage industry consolidated, the scrap industry went through another round of consolidation. Escalating costs had driven family-owned firms to sell to corporations as early as the 1950s, long before the passage of Superfund, but the 1980s and 1990s saw unprecedented consolidation of the ferrous scrap industry. Small firms operated throughout the period, though the trend of buyouts and mergers established over the previous two decades continued in the 1970s as operating costs continued their steady rise.[64] Morton Plant, then the incoming president of ISIS, predicted in 1980 further consolidation in the industry, with medium and smaller companies merging or being bought out by larger companies.[65] Plant's prediction forecasted a wave of mergers and sales over the last quarter of the century; many firms were purchased by international holding companies such as SHV Holdings NV of Utrecht, The Netherlands, which bought the David J. Joseph Company in 1975. Phillips, a Toronto-based holding company, purchased Ogden's Luria subsidiary and dozens of other scrap brokerage and processing concerns in the mid-1990s.[66]

Such transactions indicated how much the industry's structure had changed over the course of the twentieth century. Firms founded by European immigrants at the end of the nineteenth century were sold for millions of dollars to large corporations (including European corporations) by the end of the twentieth century. Large, stable brokers and dealers possessing millions of dollars of equipment populated the highest levels of the industry. The days when a Sigmund Dringer could rise from immigrant peddler to one of the larger scrap dealers in the nation within a decade were replaced by decades of stable market dominance by established firms, several more than a century old. Immigrants coming from the Caribbean and Latin America continued to enter the industry at its lowest levels, but they faced the obstacles of well-established firms with decades of connections to customers and suppliers, as well as advantages in operating capital and processing technology. Scrap had, despite all its conflicts, become part of the nation's industrial establishment.

Green or Brown?

Once recycling became an environmental virtue, scrap dealers faced a paradox. What they had been reviled for decades for doing now was done by mainstream society and celebrated as a virtue. Public collection efforts meant many dealers had new sources to purchase and in turn process and sell to manufacturers. (Some dealers were also threatened by public facilities selling directly to manufacturers, but this was not a conspicuous problem voiced by dealers in their trade publications or public announcements.) The growing market by environmentally conscious consumers to purchase recycled materials bolstered the markets for many materials. Demand for recycled paper products rejuvenated the paper market that had been in decline for decades. In many ways, the new recycling ethos represented America catching up with its scrap dealers.

Yet the environmental concerns that expanded the markets for scrap dealers also produced new tensions and recriminations. Scrap dealers had long been accused of endangering public health with germ-laden rags, noxious fumes, and disorder. Concern over toxins and environmental health crystallized in the 1960s. At the same time lawmakers recognized the importance of recycling to reduce waste streams, they also moved to reduce the impact of toxins in the environment. The original recyclers provided a public benefit, but they continued to endanger public health, now articulated as releasing toxins into the environment.

Scrap dealers lobbied Congress for almost twenty years to reduce the burdens of Superfund on scrap firms. In 1999, Congress amended the law to lower firms' liabilities. Yet the new burdens of environmental regulation reinforced the perception among many veteran scrap dealers that the federal government did not like them and was trying to drive them out of business, a perception that predated environmental regulations to the time when dealers voiced such sentiments against the War Production Board during World War II.

Federal environmental regulations and their effects brought a new complexity to the scrap industry. Scrap dealers associated environmental regulations with a new set of costs, but some realized the protections also brought opportunities. Public recycling programs increased the amounts of secondary materials recovered and sold in the 1970s. Secondary aluminum production in the United States, bolstered by container recycling laws, increased from 752,000 metric tons in 1965 to 1,577,000 metric tons in 1980. Secondary lead production increased from 450,000 metric tons to 581,000 metric tons over the same period, in part due to an automobile battery recycling program offering incentives for reclamation.[67] In addition, some scrap firms branched out into new, lucrative waste-management operations such as asbestos abatement and hazardous material storage.[68] The new regulatory environment made scrap processing and storage more expensive at the same time it facilitated increased recycling and economic opportunities handling wastes.

New regulations contributed to escalating operating costs for processors, though they were not the only causes, nor were rising costs new phenomena after 1965. Rising costs due to labor, technology, material, and transportation continued a trend in place since World War II. Battelle's survey noted continued concerns about the rising costs of operations, including "the rising minimum wage levels, increased imports of both materials and end-products, and increased transportation costs," noting that "since labor costs are increasingly difficult to control there is an even greater need on the part of secondary materials processors to find mechanized processes for handling, sorting, and upgrading raw wastes to usable products."[69]

Battelle identified several problems affecting recycling of commodities. Of particular concern to dealers were issues with customers, the federal government, new technology, and costs. One problem was what scrap firms viewed as irrational customer specifications and discriminatory government procurement procedures. Some government specifications called for use of only primary materials. Some specifications were designed to make it difficult for recycled materials to meet them. The effects of irrational specifications were reduced markets and recycle rates.[70]

Dealers cited costs and inadequate knowledge of equipment as major problems. Due to the industry's cost structure, attention to the use and purchase of the most effective equipment for firms was not a high priority for many firms. Without knowledge or competent advice, firms often made irrational equipment purchases, resulting in higher investment and maintenance costs and suboptimal productivity. Battelle's report attributed these problems to a lack of engineering expertise within the industry, reluctance to seek advice from outside experts, little exchange of information between industry members, and distrust between scrap firms and equipment manufacturers.

The availability of equipment limited firms' productivity. Scrap processing yards often kept the same technology they had used for decades. As larger competitors began investing in new machinery, such as automobile shredders, and postconsumer scrap became more complex and difficult to sort, the limits of older equipment became apparent. In many cases, equipment was too inflexible for general use or was unavailable or inadequate for firms' needs. Equipment problems often prevented processing of materials or necessitated use of manual labor, increasing labor costs, including wages and insurance, when equipment was unsuited to tasks.[71]

Federal economic policies also produced complaints from scrap processors. In the postwar period, mining companies received incentives, such as an allowance deduction from taxable income, in the tax codes to extract more materials. Scrap firms complained that the tax codes created an unfair advantage for sellers of primary materials and encouraged sale of primary materials over secondary materials, thus having a strong negative effect on recycling. Throughout the 1970s, ISIS lobbied to win tax codes that would benefit the industry, in addition to freight rates that would encourage transportation of scrap within the United States.[72]

Despite complaints concerning the costs of regulation and technology, the period between 1965 and 2000 brought scrap dealers much to celebrate. At the same time costs in the scrap industry increased, scrap sales in the United States were also on the rise. Ferrous scrap consumption peaked in the mid-1960s; over ninety million short tons of ferrous scrap were consumed in 1965. Simultaneously, scrap consumption declined steadily; in 1980 American industry consumed just fewer than seventy-six million short tons of ferrous scrap. As the overall consumption of scrap iron and steel declined, ferrous scrap sales (as measured in receipts of purchased scrap) rose from just under thirty-six million metric tons sold in 1965 to over forty-two million metric tons sold in 1980 and sixty-eight million tons sold in 2000. During a period when producing technologies changed, federal regulations increased, and concern for the environment prompted new attention toward recycling, ferrous scrap firms increased their sales.[73]

Most of the volume of scrap traded at the end of the 1970s went through large brokers, and most of the firms in the industry were small yards. The average yard had about ten employees working for near-minimum wage, often, as in the past, residents of nearby neighbor-hoods. Precise numbers and demographics of scrap employees re-mained difficult to determine, as many of the employees were part-timers or seasonal workers who might not show up in the cen-sus.[74] African Americans and recent immigrants (including Domini-can immigrants collecting materials with vans in New York City) continued to work as yard laborers and scavengers collecting small amounts from local sources for the yards. The ability to sort materi-als continued to be important; workers who could identify and sort different alloys were particularly appreciated.[75]

Other dealers recognized opportunities in the complex world of environmental health and fears. Some got into the business of haz-ardous waste removal, with asbestos removal cited as being particu-larly lucrative. Hazardous waste abatement joined the transition from scrap yards to minimills as ways in which dealers could transform their businesses to reflect new technologies, concerns, and opportu-nities. But within the industry, environmental regulation was cited as a reason why dealers could not meet the expenses to stay in business and as a cause for the decline of the small, family-owned firm.

Salvaging Recycling

The scrap industry embraced the rhetoric of recy-cling at a time when interest in recycling postconsumer materials was greater than it had been at any time since the scrap drives of World War II. Whereas the interest in the latter drives was to conserve mate-rials for short-term war production, the new ethos of recycling was meant to reduce waste's effects on the environment. Just as it had done in the 1920s, the scrap industry couched its rhetoric to position its activities as addressing the concerns of the times.

These activities could well be within the mainstream of popular environmental thought; they could make popular environmental ac-tivities possible. The criticism of New York City's suspension of glass and plastic curbside collections did not abate in 2002, and officials began to consider how to resume a popular program. The city enter-tained bids by private firms to see how much the firms would charge the city to cart away recyclables, hoping that the low bidder would charge less than what it had paid when the decision to suspend pick-ups was reached. The city had paid WMX to cart away recyclables, as New York City, like many cities in the 1990s, outsourced its recycling pickups with its garbage pickups to the same firm. Yet as the city found out as it investigated whether to resume plastic and glass pickups in the outcry that greeted the suspension of those programs, using a scrap dealer rather than WMX would generate, rather than cost, money for plastic and metal. The *New York Times* observed that the difference is that most waste companies make their profits by disposing of trash in the most efficient way possible, usually in landfills they have purchased around the country, while scrap companies attempt to squeeze out the last possible percentage of salvage from the cars and other items they process, and any excess that cannot be used and must be thrown out represents a failure and a loss.

Commissioner Doherty said at a city council hearing in January 2003 that the city would sign a contract with a large New Jersey recycling company, Hugo Neu Schnitzer East, that offered to pay the city $5.15 per ton for plastic and metal garbage, instead of charging for the service. Other, bigger garbage firms, he said, wanted to charge the city more that $67.00 per ton for removing plastic and metal refuse. The Department of Sanitation resumed collection of glass and plastic in April 2004.[76] Mark A. Izeman of the Natural Resources Defense Council, a New York–based conservation group, remarked that the city's reliance upon the scrap firm's willingness to pay for collected scrap—something scrap firms had done in the city for over a century—represented one of the most significant solid-waste developments in New York in years.[77] That this new development relied upon the century-old industrial practices of (often reviled) scrap firms revealed the gap between environmentalists and the original recyclers.

Now, in the early twenty-first century, the idea of recycling has become attractive to citizens, politicians, and engineers interested in reducing the amount of waste going into landfills. The practice of returning postconsumer and postindustrial materials to industrial production, however, involves activities that are not universally regarded as environmentally sustainable. If scrap dealers are, as they claim, original recyclers, they are also subject to new scrutiny on environmental grounds. Concerns over hazardous waste have led to new regulation, extending the image of scrap handlers as dirty to include the changing environmental concerns in the wake of *Silent Spring*.

The renewed image of scrap firms as polluters obscures their actions as recyclers—and obscures the full measure of how recycling works. Public recycling efforts often bypass scrap dealers in favor of contracting to garbage haulers, increasing costs to municipalities. The suspension and subsequent reinstatement of New York City's recycling program in league with Hugo Neu offers an opportunity to reconsider the centrality of scrap handlers in fulfilling the goals of recycling advocates. That reconsideration is long overdue.

Conclusion

If the developments of the past two centuries provide any continuity with the future, Americans' conflicting notions of what waste is will continue to foster tensions in how we use and reuse materials. Americans want to consume with no limit and to have clean homes and bodies. Americans also desire clean air, water, and landscapes.

A consumer culture defining waste as filth produces the expulsion of more matter out of American homes on the grounds that old objects are less sanitary and inferior to new objects. Industrial definition of waste as inefficiency produces the demand for affordable materials, including those found in the refuse of consumers (as well as industry, demolition sites, military ordnance, and other sources). Scrap dealers have taken advantage of the different definitions of waste since the United States industrialized to develop a commodities market for materials some deemed worthless and others coveted. In the sense that scrap dealers began collecting, processing, and selling salvaged materials in large volumes in the middle of the nineteenth century, scrap dealers may lay claim to having done the work of recycling a century before environmentalists identified the practice as a responsible way to limit pollution.

Identifying scrap processing as environmentally sound, or even a recognized public virtue, however, ignores complexities that have produced problems for scrap firms. Urban residents found scrap processing a nuisance, and municipalities formalized efforts to limit the public's exposure to waste handling. Unease with waste allowed immigrants the opportunity to establish businesses in the growing and profitable, yet stigmatized, trade. Unease with waste also pushed scrap into the backyards of those who had the least political and economic power to keep it out. The ways in which American society has defined waste have shaped our choices in how to manage wastes, shaped the industry of salvaging from wastes, and shaped the effects consumption and waste management have had on the environment and society.

Continuity and Change

These dynamics produce a marked contrast with material use and reuse from colonial times. On his long ride into the twenty-first century, Paul Revere would find much that is unfamiliar about our ways of reusing materials, but he would recognize some aspects from his own time. The idea of manufacturers seeking old materials because they are more affordable has not changed. Old metals remain prized, precious metals for their scarcity and ferrous metals and aluminum for their wide utility. Revere would be amazed at the volume of materials collected and processed in industrial production, but he would find the general concept familiar. However, a contemporary observer would have to explain to him why Americans recycle, for he would not know the idea of saving the environment by sorting discards for other people's use.

Revere's perception of material reuse would make it easier for him to understand the private scrap industry than to understand curbside pickup programs. The scrap trade is predicated on the use and reuse of waste materials. Scrap dealers depend on the demand for certain wastes by industries, including the steel, copper, aluminum, rubber, and paper industries, searching for sources of affordable raw materials. Only those wasted materials desired by manufacturers, including the materials mentioned above and wool and cotton rags used in paper manufacture, are collected and sold by dealers, creating a closed loop of materials reclamation and reuse. Mills seek out recyclable materials by salvaging scrap generated in their own production (home scrap) and by acquiring scrap from outside sources such as independent scrap dealers (purchased scrap). Scrap dealers constitute an important part of recycling, acquiring postconsumer and industrial scrap, processing the materials, and selling them to mills for reuse in industrial production. Rathje and Murphy recognize that this system of enterprise is what makes curbside collection programs work, for "recycling does not get done because it is a good thing; it gets done because it is a profitable thing, and profitability in this case depends primarily on the demand for recyclable materials."[1] This demand can be volatile—prices fluctuate from week to week and year to year. This was true in 1870 and remains true today.

The system of enterprise that emerged in the late nineteenth century allowed immigrants to establish new businesses trading waste materials. Between the Civil War and World War I, scrap provided opportunities to thousands of newcomers entering the trade as peddlers and scavengers, many of whom experienced substantial economic mobility. Though the businesses formed by some immigrants at the turn of the century established relationships with customers that made them dominant in the market and reduced opportunities for subsequent entrepreneurs (along with rising labor and equipment costs after World War II), scrap remained an industry shaped by immigrant families, and despite increased corporate ownership since the 1950s, many firms today are in their fourth or fifth generation of family ownership.

The final third of the twentieth century saw major shifts in the cultural and political forces affecting the reuse of materials in American society. The change in emphasis from conserving resources to protecting the environment altered the material reuse practices in both the private scrap industry and society at large. The new emphasis on environmental safety brought a changed focus to the concept of recycling, both in the public consciousness and in the term's centrality to the scrap industry. The vital agents of conservation had become the original recyclers, now providing an environmental service but still evaluating the discards of American consumption for profit.

By the beginning of the twenty-first century, material reuse practices had evolved so that both private and public interests were involved in reclaiming materials, and frequently private and public entities had to work together to ensure materials completed their life cycles. Many Americans expected recycling programs to exist as a matter of moral responsibility, and when programs were imperiled, residents complained. As these very public concerns about recycling and the environment played out, less conspicuous tensions between recycling and environmental health produced new challenges for recyclers. Though curbside pickup programs were common in middle-class and upper-class residential neighborhoods, the work of transporting and processing consumer recyclables was done either in industrial areas or in lower-income neighborhoods where the burdens of noise, odor, and exposure to harmful substances fell on the residents and employees of those areas.[2]

Americans now consider recycling a good thing to do for the environment, yet Americans also remain highly consumptive and dispose of more packaging materials and consumer goods than any society on Earth. The dominant image of reuse is placing recyclables in a container and having someone else take them away. What happens to the discards next is out of sight and out of mind.

It's a Dirty Job, but Someone's Gotta Do It

The complex relationship Americans have with their waste handling extends to the people charged with handling those wastes. The importance of waste trade workers, especially when there is a garbage hauler's strike, is occasionally recognized. Without their daily labors, the orderly management of wastes breaks down. The waste trades have rarely garnered respect, despite their importance. To call waste trade labor dirty work is not novel, for waste handling involves unsanitary conditions. Stuart Perry's characterization of San Francisco garbage haulers as doing dirty work defines the labor as low status and high risk and thus dirty, but also honorable and necessary. Perry's study rightly emphasizes the importance of garbage hauling, but it does not examine the many ways in which garbage haulers and other waste workers are seen as dirty.[3]

The term "dirty work" may be expanded to include other connotations of dirty behavior. The dirty connotations of the waste trades extend beyond issues of sanitation to include notions of morality and xenophobia. Garbage hauling, toxic waste storage, and scrap dealing are three waste industries whose images are associated with crime; not coincidentally, they are also strongly associated with ethnic groups who migrated to the United States in large numbers between 1870 and World War I. Contemporary examples abound in popular culture: Anthony Soprano of HBO's *The Sopranos* uses a garbage-hauling business as a front for his mafia operations, mirroring reports of toxic waste handlers being prosecuted for mafia-related activities over the past two decades. The most recent films in the *Star Wars* series feature a scheming, hook-nosed, seemingly Semitic junk dealer who enslaves the young boy who grows up to be Darth Vader. The waste trades are linked to identifiable ethnic and criminal identities in popular culture, a link that—justified or not—is both strong and enduring in American culture.[4]

The United States is by no means unique in attaching stigma to waste handlers. The work by definition occurs on the boundaries of what societies deem valuable and worthless. The industrialized world has several examples of this dynamic over the past two centuries; Donald Reid has observed the complex and often negative public image of Parisian cesspool workers despite the importance of their work. The United States, however, has hosted a dynamic conflating waste not only with ethics and social standing but also with xenophobia, raising questions of what handling this material has to do with being an American. This dimension to American definitions of waste affect past and present attitudes toward consumption and incomplete perceptions of recycling that value consumer separation yet ignore or stigmatize scrap processing.[5]

If handling wastes runs counter to American values, and yet someone handling waste is crucial to maintain American standards of hygiene and consumption, where is there room for the thousands of people working with wastes in this society? The people doing this work, like scavengers struggling to eke out a living in preindustrial societies, have tended to be newcomers and outcasts. Most were new immigrants, and many who searched for scraps lacked networks or means to find other occupations that did not require such unpleasant activities. Yet portraying the scrap peddlers and dealers who canvassed city streets as passive, exploited cogs in the machine of heavy industry misses crucial dimensions of the trade. Thousands of scrap dealers entered the trade not because they lacked means to do anything else; many could have become workers in factories. Rather, scrap dealing offered the opportunity to start a small enterprise with low expenditures. Scrap offered the autonomy of ownership—along with its risks—to newcomers, some who used the trade to build fortunes. Others used the trade as a stepping-stone to own other businesses. Scrap collection offered opportunities that fit the values and goals of many new Americans. Hundreds of these businesses have become institutions in American business, with the owners finding fortune and the firms in operation for more than a century.

If waste produces wealth, it is not without a price. Scrap firms were regarded as burdens, and the people engaged in the trade were seen as both unsanitary and morally degraded. The trade endured—

and grew—because of the wants of industry and production of ever-more scrap, but the industry never shook the negative image constructed a century ago. One of the results has been the pushing of scrap yards out of the backyards of prosperous urban neighborhoods, away from the highways of suburbia, and into the backyards of communities with little political or economic power. Since the end of World War I, most of these communities have been of people of color.

Despite cultural changes that associated scrap with patriotism during World War II and environmentally responsible behavior in the late twentieth century, scrap retains much of this stigma today, and the environmental inequalities of the early twentieth century endure at the beginning of the twenty-first century. The markets for scrap materials remain vibrant, but concerns over the processes to extract these materials from the waste stream have led to new worries over the materials left behind. Complex consumer and industrial products contain more hazards, and the scrap industry bears the burden of disposing of these feared materials. The environmental consequences of processing hazardous substances are feared, and scrap firms have borne new stigma as Superfund sites and brownfields since the early 1980s. Certainly those firms that have poorly processed or secured hazards should be held accountable for endangering the environment. Yet blaming scrap dealers for pollution ignores our larger responsibilities as consumers for producing the context for these problems. Americans fail to see how everyone is responsible for the environmental inequalities in the scrap industry, and they fail to take responsibility for producing the conditions that make scrap handling objectionable.

The Tragedy of Waste

Modern recycling continues the culture of disposability that took hold in the late nineteenth century; however, the history of reuse may inform more sustainable waste management strategies in the future. The notion of throwing away things en masse produces new problems today, as the kinds of materials discarded in complex manufactured items such as the automobile contain substances recognized as hazardous when processed or discarded. The materials put through the cycle of production are recognized as harmful, and this recognition produces regulations intended to mitigate harmful effects. In the past, issues of nuisance, aesthetic blight, and public health have spurred regulation of waste handling. For much of the industrial history of this nation, regulating the unpleasant effects of society's wastes has focused on the handlers of wastes, either disposal agents at the end stage or recycling agents preparing materials for reintroduction to the cycle of production. This focus reflects a perception that these people are a problem and threat to society, though they perform services crucial to perpetuating the consumption patterns of society.

Design for recycling may alleviate some of the problems inherent in processing scrap materials. Removing toxic or unusable materials from complex manufactured products such as automobiles and appliances would reduce the issues of hazardous materials escaping into the air, ground, and water once the products pass from consumers to recycling agents.[6] Engineering solutions are necessary but not sufficient; industrial ecologists in the early twenty-first century may have similar impacts to Henry Ford's attempts at salvage in the early twentieth century. Cultural changes in the ways Americans perceive their objects and the people who handle them are required to address environmental inequalities. American patterns of consumption and waste are deeply rooted, and attempts to develop sustainable use of resources must take the cultural assumptions regarding waste into account in order to succeed. Proposals ranging from dramatic reductions in consumption to redesigning industrial processes to eliminate hazardous byproducts are frequent, but as yet they have not substantially altered the mass production of complex wastes, nor have they alleviated the social inequalities of waste management.

T. J. Jackson Lears titled a book about the evolution of sales and advertising in American life in the early twentieth century *Fables of Abundance.*[7] That title is appropriate for an interpretation of how economic development occurred in the United States by pushing necessary yet unpleasant activity to the margins of society by believing that unfettered production and consumption did not rely on the aesthetically unpleasant work of those who traded cash for trash. Notions of class, race, ethnicity, and citizenship produced and strengthened a hegemony that built economic power by the established, legitimate institutions within society, at the expense of the marginalized people who do the work that makes abundance possible. This dynamic is complex, as individual firms and families may gain significant economic power, yet the work itself remains marginalized in American culture and geography.

Perhaps most baffling of all is the idea that people who play an important role in an activity many see as a civic virtue could be so stigmatized. Yet the idea of recycling as a sustainable environmental goal is a fantasy if the work of scrap recycling workers and firms cannot be reconciled with the accepted values and norms of industrial society. For too long the tragedy of waste in the United States has not been the inefficiency Stuart Chase warned of, but a consumptive society unwilling to look at the consequences of its actions and too willing to foist those consequences on a few, all the while failing to reconcile the desire for purity with the dangerous work that makes an orderly society possible.

Notes

INTRODUCTION

1. On efficiency measures in the petroleum industry, see Hugh S. Gorman, *Redefining Efficiency: Pollution Concerns, Regulatory Mechanisms, and Technological Change in the US Petroleum Industry* (Akron: University of Akron Press, 2001).
2. Esther Forbes, *Paul Revere and the World He Lived In* (New York: Houghton Mifflin Co., 1969).
3. Suellen M. Hoy and Michael C. Robinson's *Recovering the Past: A Handbook of Community Recycling Programs, 1890–1945* (Chicago: Public Works Historical Society, 1979) gives an excellent overview of community collection efforts dating from Salvation Army charity drives in the mid-nineteenth century to World War II scrap drives coordinated by the federal government.
4. Mary Douglas, *Purity and Danger: An Analysis of Concepts of Pollution and Taboo* (New York: Frederick A. Praeger Publishers, 1966), 36.
5. Michael Thompson, *Rubbish Theory: The Creation and Destruction of Value* (Oxford: Oxford University Press, 1979), 48.
6. Suellen Hoy, *Chasing Dirt: The American Pursuit of Cleanliness* (New York: Oxford University Press, 1995).
7. Stuart Chase, *The Tragedy of Waste* (New York: Macmillan Co., 1927). On the rise of managed corporations, see Alfred D. Chandler Jr., *The Visible Hand: The Managerial Revolution in American Business* (Cambridge, MA: Belknap Press, 1977).
8. Roger D. Waldinger, *Through the Eye of the Needle: Immigrants and Enterprise in New York's Garment Trades* (New York: New York University Press, 1986); Roger Waldinger, Howard Aldrich, Robin Ward, et al., *Ethnic Entrepreneurs: Immigrant Business in Industrial Societies* (Newbury Park, CA: Sage Publications, 1990); Ivan Light and Steven J. Gold, *Ethnic Economies* (San Diego: Academic Press, 2000). Middleman minority studies include Edna Bonacich, "Middleman Minorities and Advanced Capitalism,"

Ethnic Groups 2 (1980): 211–19; Walter P. Zenner, "Introduction: Symposium on Economics and Ethnicity: The Case of the Middleman Minorities," *Ethnic Groups* 2 (1980): 185–87.

9. Philip B. Scranton, *Endless Novelty: Specialty Production and American Industrialization, 1865–1925* (Princeton, NJ: Princeton University Press, 1997), 355.

10. David Naguib Pellow's case study on recycling centers in Chicago highlights these environmental inequalities. David Naguib Pellow, *Garbage Wars: The Struggle for Environmental Justice in Chicago* (Cambridge, MA: MIT Press, 2002).

11. See Craig McMillan Richmond, "Simulating Differences in Ferrous Scrap Prices Over Geographic Space Using the Logistic Model of Choice for Differentiated Products" (Ph.D. diss., University of Pittsburgh, 1997), for an econometric study of ferrous scrap markets.

12. Jennifer Carless, *Taking out the Trash: A No-Nonsense Guide to Recycling* (Washington, DC: Island Press, 1992); Debi Kimball, *Recycling in America: A Reference Handbook* (Santa Barbara, CA: ABC-CLIO, 1992); Adam S. Weinberg, David N. Pellow, and Allan Schnaiberg, *Urban Recycling and the Search for Sustainable Community Development* (Princeton, NJ: Princeton University Press, 2000).

13. John Tierney, "Recycling Is Garbage," *New York Times Magazine*, 30 June 1996, 24–29, 44, 48–49, 53.

14. Martin V. Melosi, *The Sanitary City: Urban Infrastructure in America from Colonial Times to the Present* (Baltimore: Johns Hopkins University Press, 2000), 413.

15. William Rathje and Cullen Murphy, *Rubbish! The Archaeology of Garbage* (New York: HarperCollins, 1992), 203.

16. David T. Allen and Nasrin Behmanesh, "Wastes as Raw Materials," in *The Greening of Industrial Ecosystems,* ed. Braden R. Allenby and Deanna J. Richards (Washington, DC: National Academy Press, 1994), 69–89.

17. Christine Meisner Rosen, "Industrial Ecology and the Greening of Business History," *Business and Economic History* 26 (fall 1997): 123–37; Pierre Desrouchers, "Market Processes and the Closing of 'Industrial Loops': A Historical Reappraisal," *Journal of Industrial Ecology* 4, no. 1 (winter 2000): 29–43.

18. George H. Thurston, *Pittsburgh and Allegheny in the Centennial Year* (Pittsburgh: A. A. Anderson & Son, 1876), 185.

19. Susan Strasser, *Waste and Want: A Social History of Trash* (New York: Metropolitan Books, 1999), 12–16; Hoy, *Chasing Dirt.* Christine Stansell, Eric Schneider, and David Nasaw have examined scavenging as a strategy of impoverished urban women and children. David Nasaw, *Children of the City: At Work and at Play* (Garden City, NY: Anchor Press/Doubleday, 1985); Christine Stansell, *City of Women: Sex and Class in New York, 1789–1860* (Urbana: University of Illinois Press, 1987); Eric Schneider, *In the Web of Class: Delinquents and Reformers in Boston, 1810s–1930s* (New York: New York University Press, 1992).

20. Martin V. Melosi, *Garbage in the Cities: Refuse, Reform, and the Environment, 1880–1980* (College Station: Texas A&M University Press, 1981); Melosi, *The Sanitary City*; Joel A. Tarr, *The Search for the Ultimate Sink: Urban Pollution in Historical Perspective* (Akron, OH: University of Akron Press, 1996).

21. For a spatial analysis of the proximity of scrap metal firms to customers, see John William Maher, "Retrieving the Obsolete: Formation of the American Scrap Steel Industry, 1870–1933" (Ph.D. diss., University of Maryland, 1999).

CHAPTER 1 RAGS AND OLD IRON

1. Letter to Peter Dixon, 20 September 1872, Francis Bannerman Son, Inc., outbound letters, Accession 2185, ser. 3, box 1, vol. 119, Hagley Museum and Library, Wilmington, DE.
2. Laurence Fontaine, *History of Pedlars in Europe,* trans. Vicki Whittaker (Durham, NC: Duke University Press, 1996); Betty Naggar, "Old-Clothes Men: 18th and 19th Centuries," *Jewish Historical Studies* 31 (1988–90): 171–91.
3. Letter to Peter Dixon, 20 September 1872, Francis Bannerman Son, Inc., outbound letters, Accession 2185, ser. 3, box 1, vol. 119, Hagley Museum and Library, Wilmington, DE.
4. Strasser, *Waste and Want,* 21–67.
5. Betty Naggar, *Jewish Pedlars and Hawkers, 1740–1940* (Camberley, England: Porphyrogenitus, 1992); Edwin C. Barringer, *The Story of Scrap* (Washington, DC: ISIS, 1954): 73; Charles H. Lipsett, *100 Years of Recycling History: From Yankee Tincart Peddlers to Wall Street Scrap Giants* (New York: Atlas Publishing Co., 1974), 12.
6. Strasser, *Waste and Want,* 100; Barringer, *The Story of Scrap,* 73.
7. Strasser, *Waste and Want,* 100.
8. Ibid., 98; Barter list, 1854, box 5; Noyes MSS, Baker Library, Harvard Business School.
9. Strasser, *Waste and Want,* 99; William Gilbert Lathrop, *The Brass Industry in the United States: A Study of the Origin and Development of the Brass Industry in the Naugatuck Valley and Its Subsequent Extension over the Nation* (Mount Caramel, CT: W. G. Lathrop, 1926), 37–38, 77–79.
10. Strasser, *Waste and Want,* 101–2.
11. *Pennsylvania Gazette,* 4 March 1731; 3 November 1737; 20 April 1732.
12. Carole Shammas, *The Pre-Industrial Consumer in England and America* (New York: Clarendon Press, 1990). Susan Strasser observes a strong gender dimension to this trade. Housewives would trade rags for consumer goods or "rag money" to the male peddlers who went door to door. Strasser, *Waste and Want,* 69–110.
13. Consumers often did not discard textiles immediately after the original product had served its use. American women reused their old clothes, claiming what Strasser calls a "stewardship of objects," by reusing them in quilts, upholstery, or other household objects. Strasser, *Waste and Want,* 21–67.
14. "Importance of Rags," *Scientific American,* n.s., 15 (28 July 1866): 71.
15. Judith A. McGaw, *Most Wonderful Machine: Mechanization and Social Change in Berkshire Paper Making, 1801–1885* (Princeton, NJ: Princeton University Press, 1987): 39–40.
16. A detailed description of the Fourdinier's effect on paper making is Robert H. Clapperton, *The Paper-Making Machine* (New York: Pergamon Press, 1967), 45–53, 115–253. Also see McGaw, *Most Wonderful Machine,* 97–98.
17. McGaw, *Most Wonderful Machine,* 40.

18. Mansel G. Blackford and K. Austin Kerr, *Business Enterprise in American History* (Boston: Houghton Mifflin, 1986), 85.
19. Lillian R. Greenstein, "The Peddlers of Bay City," *Michigan Jewish History* 25, nos. 1–2 (1985): 10–17. Louis Schmier's article dealing primarily with race relations has illuminating information on old and new material exchanges between peddlers and customers. Louis Schmier, "'For Him the "Schwartzers" Couldn't Do Enough': A Jewish Peddler and His Black Customers Look at Each Other," *American Jewish History* 73, no. 1 (1983): 39–55.
20. McGaw, *Most Wonderful Machine*, 40.
21. Ibid., 40–41.
22. Ibid., 191.
23. Strasser, *Waste and Want*, 85; McGaw *Most Wonderful Machine*, 191–92; "Parliamentary Reports on the Rag Trade of Foreign Countries," *Practical Magazine*, n.s., 5 (1875): 221.
24. Strasser, *Waste and Want*, 85.
25. McGaw, *Most Wonderful Machine*, 194.
26. Ibid., 191–92; "American Industries," no. 72, 275. *Scientific American* 44 (30 April 1881): 271, 275–76; *Delaware County American,* 18 October 1865.
27. Yukichi Fukuzawa, *The Autobiography of Yukichi Fukuzawa*, trans. Eiichi Kiyooka (New York: Columbia University Press, 1966), 110–17.
28. Several trade publications stressed the need to collect scrap materials at the turn of the century. See "Treasures of a Scrap Pile," *Scientific American* 77 (18 December 1897): 390; H. Hill, "Traffic in Waste Paper," *Municipal Affairs* 2 (June 1898): 134–44; "Recovered Rubber," *Scientific American* 54 (19 July 1902): 222; "Recovery of Waste in Using White Metals," *Scientific American* 87 (8 November 1902): 305; "Wastes of a Blast Furnace: How They Are Utilized," *Scientific American* 71 (10 June 1911): 355–56.
29. Data provided by PSMedia's City Directories Online site, now defunct. Directories cited include *Longworth's American Almanac, New York, New York Register and City Directory for 1840* (New York: Longworth, 1840); *Doggett's New York City, New York Street Directory for 1851* (New York: Doggett, 1851); and *Wilson's Business Directory of New York City, New York for 1860* (New York: Wilson, 1860).
30. *A. [McElroy's] Philadelphia, Pennsylvania General and Business Directory for 1840* (Philadelphia: A. McElroy, 1840); *A. McElroy's Philadelphia, Pennsylvania General and Business Directory for 1851* (Philadelphia: A. McElroy, 1851); *Gopsill's Philadelphia, Pennsylvania General and Business Directory for 1860* (Philadelphia: Gopsill, 1860).
31. Morillo Noyes MSS, box 5, no. 778/1859–1877/N956, Baker Library, Harvard Business School. Quoted in Strasser, *Waste and Want*, 75–76, 103.
32. Harold S. Wilson, *Confederate Industry: Manufacturers and Quartermasters in the Civil War* (Jackson: University Press of Mississippi, 2002), 133–40.
33. "Importance of Rags," *Scientific American*, n.s., 15 (28 July 1866): 71.
34. William T. Hogan, *Economic History of the Iron and Steel Industry in the United States* (Lexington, MA: Heath, 1971), 31–38.
35. Sir Ronald Prain, *Copper: The Anatomy of an Industry* (London: Mining Journal Books, 1975), 70.
36. Charles K. Hyde, *Copper for America: The United States Copper Indus-

try from Colonial Times to the 1990s (Tucson: University of Arizona Press, 1998); Prain, *Copper*, 70.

37. John M. Ball, *Reclaimed Rubber: The Story of an American Raw Material* (New York: Rubber Reclaimers Association, 1947), 25.
38. Ball, *Reclaimed Rubber*, 44–71.
39. Strasser, *Waste and Want*, 103; Howard Wolf, *The Story of Scrap Rubber* (Akron, OH: A. Schulman, 1943), 19–33.
40. Strasser, *Waste and Want*, 103.
41. "Saving Old Rubber," *Manufacturer and Builder* 2 (June 1870): 173.
42. Blackford and Kerr, *Business Enterprise in American History*, 85.
43. Lipsett, *100 Years of Recycling History*, 26.
44. Richard C. Wade, *The Urban Frontier: The Rise of Western Cities, 1790–1830* (Cambridge, MA: Harvard University Press, 1959); William Cronon, *Nature's Metropolis: Chicago and the Great West* (New York: W. W. Norton, 1991).
45. *Iron Age* carried regular reports on the sale of old rails. See, for example, "Trade Report," *Iron Age*, 27 February 1873, 18.
46. "Fire in Peck-Slip," *New York Times*, 15 December 1865; "Fires," *New York Times*, 24 December, 1865; "Fires," *New York Times*, 24 March, 1866; "Fire This Morning," *New York Times*, 3 July 1872; "Three Firemen Buried in Ruins of a Falling Building," *New York Times*, 5 August 1874.
47. Stuart E. Perry, *Collecting Garbage: Dirty Work, Clean Jobs, Proud People* (New Brunswick, NJ: Transaction Publishers, 1998).
48. Scranton, *Endless Novelty*.
49. Francis Bannerman to Mr. Peter Dixon, 20 September 1872; George W. Raymond to Francis Bannerman Son, Inc., 28 March 1879; Francis Bannerman to Captain Willse, USN, 24 June 1880; Col. A. Crispin, Office of U.S. Ordnance Agency to Francis Bannerman, 20 June 1882; Francis Bannerman Son, Inc., outbound letters, Accession 2185, ser. 3, box 1, vol. 119, Hagley Museum and Library, Wilmington, DE.
50. Lincoln Diamant, "The Great Chain Hoax," *Hudson Valley Regional Review* 7, no. 1 (March 1990): 44–57.
51. *Wilson's Business Directory of New York City, New York for 1860* (New York: Wilson, 1860); *Trow's New York, New York City Directory for 1866* (New York: Trow, 1866); *Trow's New York, New York City Directory for 1880–81* (New York: Trow, 1881); *Trow's New York, New York City Directory for 1890* (New York: Trow, 1891).
52. *Gopsill's Philadelphia, Pennsylvania General and Business Directory for 1860* (Philadelphia: Gopsill, 1860); *Gopsill's Philadelphia, Pennsylvania General and Business Directory for 1870* (Philadelphia: Gopsill, 1870); *Gopsill's Philadelphia, Pennsylvania City Directory for 1880* (Philadelphia: Gopsill, 1880); *Gopsill's Philadelphia, Pennsylvania General and Business Directory for 1890* (Philadelphia: Gopsill, 1890).
53. Walter Licht, *Industrializing America: The Nineteenth Century* (Baltimore: Johns Hopkins University Press, 1995), 120–122.
54. *Scrap Age Bicentennial Edition* (Niles, IL: Three Sons Publishing Co., 1977), 37–38.
55. *Report of the Secretary of the American Iron and Steel Association* (Philadelphia: American Iron and Steel Association, 1872), 38–39.
56. *Annual Statistical Report of the American Iron and Steel Association*, (Philadelphia: American Iron and Steel Association, 1888), 16.

57. "Trade Report," *Iron Age*, 27 February 1873, 18; "Trade Report," *Iron Age*, 8 May 1873, 18; "Trade Report," *Iron Age*, 19 June 1873, 18; "Trade Report," *Iron Age*, 20 November 1873, 20; "Trade Report," *Iron Age*, 25 December 1873, 8.
58. George Rosen, *A History of Public Health*, expanded ed. (Baltimore: Johns Hopkins University Press, 1993), 168–319.
59. Woodin and Little Pump House, *Catalog No. 36* 1881[?], Smithsonian Institute Libraries, alphabetical trade catalog collection.
60. "Scrap-Iron Forgings," *Scientific American* 10 (June 1864): 377.
61. Gamaliel G. Smith et al., Appellants, v. Daniel Pettee et al., Respondents [no number in original], Court of Appeals of New York 70 NY 13; 1877 NY LEXIS 579.
62. State, Josiah Clark, Prosecutor, v. Mayor, &c., of New Brunswick, Supreme Court of New Jersey 43 NJL. 175; 1881; McDowell v. Rissell Supreme Court of Pennsylvania 37 Pa. 164; 1860 Pa. LEXIS 197, 17 October 1860, Decided; Diehl v. Holben Supreme Court of Pennsylvania 39 Pa. 213; 1861 Pa. LEXIS 184, 6 May 1861, Decided; Stephen Coleman, Plaintiff in Error, v. The People of the State of New York, Defendants in Error. Court of Appeals of New York 55 NY 81; 1873 NY LEXIS 138; Gamaliel G. Smith et al., Appellants, v. Daniel Pettee et al., Respondents. Court of Appeals of New York 70 NY 13; 1877 NY LEXIS 579; Nixon v. McCrory. Supreme Court of Pennsylvania 101 Pa. 289; 1882 Pa. LEXIS 250; Lesser v. Perkins. Supreme Court of New York, General Term, First Department 4 NYS. 53; 1889 NY Misc. LEXIS 180.
63. Christine Stansell observes that scavenging seems to have been a widespread practice among the laboring classes in the eighteenth century, but after 1820, those who could afford to probably began to desist sending children to scavenge, since child scavengers were likely to be arrested for vagrancy. After 1820, most child scavengers were from the poorest classes. Stansell, *City of Women*, 50–52.
64. *Laws and Ordinances Ordained and Established by the Mayor, Aldermen, and Commonalty of the City of New York in Common Council Convened* (New York: City of New York, 1817), 112.
65. Stansell, *City of Women*, 51.
66. On nineteenth-century nuisance law in the United States, see Horace Gay Wood, *A Practical Treatise on the Law of Nuisances in Their Various Forms* (Albany: John D. Parsons, 1875); Joseph A. Joyce and Howard C. Joyce, *Treatise on the Law Governing Nuisances: With Particular Reference to Its Application to Modern Conditions and Covering the Entire Law Relating to Public and Private Nuisances, Including Statutory and Municipal Powers and Remedies, Legal and Equitable* (Albany: M. Bender & Co., 1906).
67. Christine Rosen argues that nineteenth-century nuisance laws were enforced by judges using something akin to cost-benefit analysis, with the costs balancing the need to alleviate human discomfort with the cost to industry to prevent nuisances. Rosen found that judges in industrial Pennsylvania and New York tended to side with industry in nuisance cases. Christine Rosen, "Differing Perceptions of the Value of Pollution Abatement across Time and Place: Balancing Doctrine in Pollution Nuisance Law, 1840–1906," *Law and History Review* 2 (fall 1993): 303–81.
68. Allegations of fraud and disputes over goods are found in McDowell v. Rissell Supreme Court of Pennsylvania 37 Pa. 164; 1860 Pa. LEXIS 197,

17 October 1860, Decided; Diehl v. Holben Supreme Court of Pennsylvania 39 Pa. 213; 1861 Pa. LEXIS 184, 6 May 1861, Decided; Stephen Coleman, Plaintiff in Error, v. The People of the State of New York, Defendants in Error Court of Appeals of New York 55 NY 81; 1873 NY LEXIS 138.

CHAPTER 2 NEW AMERICAN ENTERPRISES

1. "Failure of an Erie Suit," *New York Times*, 9 October 1877.
2. Stuart M. Blumin, *The Emergence of the Middle Class: Social Experience in the American City, 1760–1900* (New York: Cambridge University Press, 1989).
3. On the rise of American department stores between 1850 and 1890, see Susan Porter Benson, *Counter Cultures: Saleswomen, Managers, and Customers in American Department Stores, 1890–1940* (Urbana: University of Illinois Press, 1986). On mail-order catalogs, see Cronon, *Nature's Metropolis*, 333–40.
4. On changing dress sense and manners over the nineteenth century, see Karen Halttunen, *Confidence Men and Painted Women: A Study of Middle-Class Culture in America, 1830–1870* (New Haven: Yale University Press, 1982); John F. Kasson, *Rudeness and Civility: Manners in Nineteenth-Century Urban America* (New York: Hill and Wang, 1990).
5. Melosi, *The Sanitary City*, 261–80.
6. Ronald C. Tobey discusses federal housing policy in regard to electrification of American homes; the same political processes also mandated running water and gas be standard in new (and newly mortgaged) residences. Ronald C. Tobey, *Technology as Freedom: The New Deal and the Electrical Modernization of the American Home* (Berkeley: University of California Press, 1996).
7. Strasser, *Waste and Want*, 38–67.
8. Hoy, *Chasing Dirt*.
9. Melosi, *The Sanitary City*, 175–212; Melosi, *Garbage in the Cities*, 181–83.
10. Samuel H. Preston, Douglas Ewbank, and Mark Hereward, "Child Mortality Differences by Ethnicity and Race in the United States: 1900–1910," in *After Ellis Island: Newcomers and Natives in the 1910 Census*, ed. Susan Cotts Watkins (New York: Russell Sage Foundation, 1994), 66–69.
11. Daniel Nelson, *Frederick W. Taylor and the Rise of Scientific Management* (Madison: University of Wisconsin Press, 1980).
12. Kenneth D. Durr and James H. Lide, "A 'New Industrial Philosophy'? World War II and the Roots of Corporate Recycling." Paper presented at American Society for Environmental History Conference, Tucson, AZ, 1999; Thomas Martin McCarthy, "The Road to Respect: Americans, Automobiles and the Environment" (Ph.D. diss., Yale University, 2001).
13. John Bodnar, *The Transplanted: A History of Immigrants in Urban America* (Bloomington: Indiana University Press, 1985): 20; Roger Daniels, *Coming to America: A History of Immigration and Ethnicity in American Life* (New York: HarperCollins, 1990); Hasia R. Diner, *A Time for Gathering: Second Migration, 1820–1880* (Baltimore: Johns Hopkins University Press, 1992); Gerald Sorin, *A Time for Building: The Third Migration, 1880–1920* (Baltimore: Johns Hopkins University Press, 1992); David Ward, *Cities and Immigrants: A Geography of Change in Nineteenth Century America* (New York: Oxford University Press, 1971), 52–53.

14. John Bodnar, Roger Simon, and Michael P. Weber, *Lives of Their Own: Blacks, Italians and Poles in Pittsburgh, 1900–1960* (Urbana: University of Illinois Press, 1982), 20.

15. Thomas Bell, *Out of This Furnace* (Boston: Little, Brown and Co., 1941); Upton Sinclair, *The Jungle* (New York: Doubleday, Page & Co., 1906).

16. Waldinger, *Through the Eye of the Needle*, 19–47.

17. Perry, *Collecting Garbage.*

18. Stansell, *City of Women*, 51.

19. Melosi, *Garbage in the Cities*, 71–72.

20. Nasaw, *Children of the City*, 88; Strasser, *Waste and Want*, 110–16.

21. Edward Schlezinger interviewed by Judy Blair recorded 14 May 1985, Columbus Jewish Historical Committee's Oral History Project (http://www.gcis.net/cjhs/Interviews/HTML/schlezinger_ed_blair1.htm).

22. Nasaw, *Children of the City*, Stansell, *City of Women*, and Strasser, *Waste and Want*, all suggest scavenging was common to recent immigrants in American cities. The Integrated Public Use Microdata Series (IPUMS–98), created at the University of Minnesota in October 1997, consists of twenty-five high-precision samples of the American population drawn from thirteen federal censuses between 1850 and 1990. The 1880 and 1920 samples include nonidentical yet comparable information on occupation and ethnicity that is instructive in recounting the demographic history of America's scrap industries. Steven Ruggles and Matthew Sobek et. al., *Integrated Public Use Microdata Series: Version 2.0* (Minneapolis: Historical Census Projects, University of Minnesota, 1997), accessed at http://www.IPUMS.umn.edu. My methods and detailed findings on the structure of the waste trades in the 1880 and 1920 IPUMS samples may be found in Carl Zimring, "Recycling for Profit: The Evolution of the American Scrap Industry" (Ph.D. diss., Carnegie Mellon University, 2002), 64–102.

23. Watkins, "Background: About the 1910 Census," 25–26.

24. For detailed analysis of waste-trade workers' demographic characteristics, see Zimring, "Recycling for Profit," 70–76, 90–100.

25. Archibald MacLeish, *Jews in America* (New York: Random House, 1936), 9. On Italians in the waste disposal industry, see Melosi, *Garbage in the Cities*, 71. Perry's study of San Francisco garbage collectors focuses on an Italian-American cooperative; Perry, *Dirty Work.*

26. Alan M. Kraut, "The Butcher, The Baker, The Pushcart Peddler: Jewish Foodways and Entrepreneurial Opportunity in the East European Immigrant Community, 1880–1940," *Journal of American Culture* 6 (1983): 71–83.

27. Oliver B. Pollak, "The Jewish Peddlers of Omaha," *Nebraska History* 63 (1982): 474–501.

28. "Receivers of Stolen Goods," *New York Times*, 6 July 1866.

29. Ibid.

30. "The Refuse of the City," *New York Times*, 4 December 1881.

31. George Waring, "The Disposal of a City's Waste," quoted in Melosi, *Garbage in the Cities*, 72.

32. Lipsett, *100 Years of Scrap Recycling History*, 190.

33. For additional anecdotes, see George Thornton Fleming, *History of Pittsburgh and Environs, from Prehistoric days to the Beginning of the American Revolution* (New York: American Historical Society, 1922): 323–24. The National Council of Jewish Women's Pittsburgh chapter conducted oral interviews with many Jewish residents, including some scrap deal-

ers. These are available at the Archives of Industrial Society in Pittsburgh. The Institute of Scrap Iron and Steel (ISIS) conducted interviews with some of its senior members in 1980, available as "The Original Recyclers" videocassette through the Institute of Scrap Recycling Industries, Washington DC.

34. Directories cited include *A. [McElroy's] Philadelphia, Pennsylvania General and Business Directory for 1840* (Philadelphia: A. McElroy, 1840); *A. McElroy's Philadelphia, Pennsylvania General and Business Directory for 1851* (Philadelphia: A. McElroy, 1851); *Gopsill's Philadelphia, Pennsylvania General and Business Directory for 1870* (Philadelphia: Gopsill, 1870); *Gopsill's Philadelphia, Pennsylvania City Directory for 1880* (Philadelphia, 1880); *Gopsill's Philadelphia, Pennsylvania General and Business Directory for 1890* (Philadelphia: Gopsill, 1890); *Gopsill's Philadelphia, Pennsylvania General and Business Directory for 1900* (Philadelphia: Gopsill, 1900); *Gopsill's Philadelphia, Pennsylvania General and Business Directory for 1910* (Philadelphia: Gopsill, 1910).

35. The 1920 figure includes thirty-six automobile salvage firms, making Detroit an early leader in automobile salvage as well as automobile production. *Detroit, Michigan City and Business Directory for 1890* (Detroit: R. L. Polk & Co., 1890); *Detroit, Michigan City and Business Directory for 1910* (Detroit: R. L. Polk & Co., 1910); *R. L. Polk & Co.'s. Detroit, Michigan General and Business Directory for 1920–21* (Detroit: R. L. Polk & Co., 1920–21).

36. *The Lakeside Chicago, Illinois General and Business Directory for 1890* (Chicago: Chicago Directory Co., 1891); *Lakeside Chicago, Illinois General Directory for 1910* (Chicago: Chicago Directory Co., 1910); *Lakeside Chicago, Illinois General Directory for 1917* (Chicago: Chicago Directory Co., 1917).

37. Fleming, *History of Pittsburgh and Environs*, 323–24.

38. Lipsett, *100 Years of Scrap Recycling History*, 22–28; "The Scrapmen," *Fortune*, January 1949, 86–91, 134–139.

39. Lipsett, *100 Years of Scrap Recycling History*, 188.

40. Mahler's comments are found on a videotape of a roundtable interview of veteran scrap dealers. *Our Heritage I*, a forty-minute tape culled from four hours of filming done June 1980, is available at the Institute of Scrap Recycling Industries, Washington, DC.

41. *Scrap Age Bicentennial Edition* (Niles, IL: Three Sons Publishing Co., 1977): 52.

42. Cox and Sons Co., *Trade Catalog* (Philadelphia: Cox and Sons Company, 1883), Smithsonian Institute Libraries; Canton Foundry and Machine Company, *Trade Catalog* (Milwaukee: Canton Foundry and Machine Company, 1920), Smithsonian Institute Libraries.

43. *Scrap Age Bicentennial Edition*, 51–52.

44. Ibid., 53–55.

45. Joe Brenner, Trading as Reliable Junk Company and the Ocean Accident and Guarantee Corporation, Ltd. vs. Toba Brenner and Mary Brenner. 127 Md. 189; 96 A. 287; 1915 Md. LEXIS 20, December 2, 1915, Decided.

46. "Economy in Scrap Yard Arrangement," *Iron Age*, 16 December 1915, 1403–4.

47. Waldinger, *Through the Eye of the Needle*, 19–47.

48. Lipsett, *100 Years of Scrap Recycling History*, 22–28.

CHAPTER 3 NUISANCE OR NECESSITY?

1. Aaron P. Levinson, *If Only Right Now Could Be Forever* (Hillsboro, OR: Media Weavers, 1987), 5.
2. Levinson, *If Only Right Now Could Be Forever*, 17.
3. "America's Richest War Bride," *Scientific American*, 24 November 1917.
4. Roland Marchand, *Advertising the American Dream: Making Way for Modernity, 1920–1940* (Berkeley: University of California Press, 1985), 156–60.
5. Lipsett, *Industrial Wastes and Salvage*, 164. For city directory data detailing the number of scrap, junk, and rag dealers in individual cities between 1840 and 1921, see Zimring, "Recycling for Profit," 33, 48, 81–82, 302–3.
6. *R. L. Polk & Co's. Detroit, Michigan General and Business Directory for 1920–21* (Detroit: R. L. Polk & Co., 1920–21); "Map Drive to Save Westchester Beauty," *New York Times*, 27 February 1929; *The Lakeside Chicago, Illinois General and Business Directory for 1890* (Chicago: Chicago Directory Co., 1891); *The Lakeside Chicago, Illinois General & Business Directory for 1917* (Chicago: Chicago Directory Co., 1917).
7. George H. Manlove, "Junk Pile Transformed into Gold," *Iron Trade Review*, 9 May 1918, 1173–76.
8. State, Josiah Clark, Prosecutor, v. Mayor, &c., of New Brunswick, Supreme Court of New Jersey 43 NJL 175; 1881; McDowell v. Rissell Supreme Court of Pennsylvania 37 Pa. 164; 1860 Pa. LEXIS 197, 17 October 1860, Decided; Diehl v. Holben Supreme Court of Pennsylvania 39 Pa. 213; 1861 Pa. LEXIS 184, 6 May 1861, Decided; Stephen Coleman, Plaintiff in Error, v. The People of the State of New York, Defendants in Error. Court of Appeals of New York 55 NY 81; 1873 NY LEXIS 138; Gamaliel G. Smith et al., Appellants, v. Daniel Pettee et al., Respondents. Court of Appeals of New York 70 NY 13; 1877 NY LEXIS 579; Nixon v. McCrory. Supreme Court of Pennsylvania 101 Pa. 289; 1882 Pa. LEXIS 250; Lesser v. Perkins. Supreme Court of New York, General Term, First Department 4 NYS. 53; 1889 NY Misc. LEXIS 180.
9. "Failure of an Erie Suit," *New York Times*, 9 October 1877.
10. Ibid.
11. W. H. Parry, "Buying and Selling Brass Foundry Scrap," *Metal Industry* 12 (June 1914): 237–38.
12. Ibid.
13. Ibid.
14. "Business and Old Iron," *Technical World Magazine*, April 1914, 244–45.
15. J. P. Alexander, "Sales of Scrap Materials," *Electric Railway Journal*, 23 January 1915, 192–93.
16. Principio Forge Co. to John S. Wirt, 16 September 1895, George P. Whitaker Co., MS 1730.1, Manuscripts Department, Maryland Historical Society Library, Baltimore.
17. Zerkow's obsession with the elusive gold plates is most explicit in chapter 12. Frank Norris, *McTeague* (New York: Doubleday and McClure, 1899).
18. *The Original Recyclers*, a videotaped interview conducted by ISIS in 1980, available from the Institute of Scrap Recycling Industries (ISRI), Washington, DC.
19. Mordecai Richler, *The Apprenticeship of Duddy Kravitz* (Middlesex, England: Penguin Books, 1964), 264–65.

20. Levinson, *If Only Right Now Could Be Forever*, 17.
21. "It Pays to Watch the Dump," *System* 31 (January 1917): 97–98.
22. Theodore Waters, "The Chemical House That Jack Built," *Cosmopolitan Magazine* 43 (July 1907): 290–93; "A Chance to Save Money from the Refuse of New York City," *Engineering News* 67 (8 February 1912): 265. For a discussion of the waste to wealth movement, see Melosi, *Garbage in the Cities*, 181–83.
23. "Reclaiming the Scrap Pile," *Industrial Management* 18 (April 1917): 734–36.
24. Lipsett, *100 Years of Scrap Recycling History*, 22.
25. Rosen, *A History of Public Health*, 270–472; Melosi, *The Sanitary City*, 58–72.
26. "Metropolitan Board of Health," *New York Times*, 27 March 1866; "The Rag Men," *New York Times*, 18 May 1872; "Fires, Perth Amboy, New Jersey: 4th Fire in 5 Days in Junk Shop of a Ruderman: Latter Arrested," *New York Times*, 17 June 1921; "Fires, New York City: Atlantic Avenue, 157: Rag Storehouse," *New York Times*, 23 July 1921.
27. One immediate physical threat was the risk of handling ammunition, which could explode and maim workers. "Accidents, Explosions: New York City: Junk Dealers Breaking Up Shells, Three Injured," *New York Times*, 29 March 1915.
28. Harry H. Grigg and George E. Haynes, *Junk Dealing and Juvenile Delinquency*, text by Albert E. Webster (Chicago: Juvenile Protective Association, [1919?]), 50.
29. Grigg and Haynes, *Junk Dealing and Juvenile Delinquency*, 50.
30. Jane Addams, "The Subtle Problems of Charity," *Atlantic Monthly*, February 1899, 163–79.
31. Court of Appeals of New York, 5 NY 81; 1873 NY Lexis 138, 25 September 1873, Argued, 18 November 1873, Decided, Courtesy Lexis-Nexus.
32. Jacob A. Riis, "The Making of Thieves in New York," *Century* 49, no. 1 (November 1894): 109–16.
33. Grigg and Haynes, *Junk Dealing and Juvenile Delinquency*, 28.
34. "City Regulation of Junk Shops," *American City*, July 1918, 24–25.
35. Grigg and Haynes, *Junk Dealing and Juvenile Delinquency*, 50.
36. See Louis Galambos's classic study of the cotton industry for an example of trade association formation. Louis Galambos, *Competition and Cooperation: The Emergence of a National Trade Association* (Baltimore: Johns Hopkins Press, 1966).
37. "Aims of a Scrap-Iron Organization," *Iron Age*, 16 July 1914, 141.
38. National Association of Waste Material Dealers, *Fifteenth Anniversary Blue Book* (New York: National Association of Waste Material Dealers, 1928).
39. Samuel P. Hays, *Conservation and the Gospel of Efficiency* (Cambridge, MA: Harvard University Press, 1959). On the adoption of conservation by American businesses in the 1920s, see Hal K. Rothman, *Saving the Planet: The American Response to the Environment in the Twentieth Century* (Chicago: Ivan R. Dee, 2000), 60–84.
40. Chase, *The Tragedy of Waste*.
41. "Dealers In Waste Reclaim Millions," *New York Times*, 17 December 1913.
42. "American Iron and Steel Institute Asked to Condemn Direct Dealing and to Make Scrap Marketing Survey," *Institute Bulletin of the Institute of Scrap Iron and Steel, Inc.* 3 (May–June 1930): 1–2.

43. "Aims of a Scrap-Iron Organization," 141.
44. U.S. Department of Commerce, *Elimination of Waste: Classification of Iron and Steel Scrap* (Washington, DC: GPO, 1926).
45. National Association of Waste Material Dealers, *Twenty-Fifth Anniversary Blue Book* (New York: National Association of Waste Material Dealers, 1938), 41.
46. "American Iron and Steel Institute Asked to Condemn Direct Dealing and to Make Scrap Marketing Survey," *Institute Bulletin of the Institute of Scrap Iron and Steel, Inc.* 3 (May–June 1930): 1–2.
47. Lipsett, *100 Years of Scrap Recycling History.*
48. Gerald A. Gutenschwager, "The Scrap Iron and Steel Industry in Metropolitan Chicago" (Ph.D. diss., University of Chicago, 1957), 57–70.
49. Oliver B. Pollak argues that pressure for second-generation immigrants to find more respectable occupations, as well as changes in the economy, made peddling a marginal activity in Omaha by 1940. Pollak, "The Jewish Peddlers of Omaha," 474–501.
50. For example, the depression of 1920–21. As iron and steel production declined in 1920, so did prices for ferrous scrap, throwing many smaller firms into ruin as their supplies stagnated for months. "Business Troubles; New York City: Eureka Scrap Iron and Metal," *New York Times*, 21 January 1921; "Business Troubles; New York City: Long Island Scrap Metal Company, Inc.," *New York Times*, 8 April 1921.
51. "American Iron and Steel Institute Asked to Condemn Direct Dealing," 1–2.
52. U.S. Congress, Senate, Committee on Military Affairs, *Hearings Before a Subcommittee of the Committee on Military Affairs on S. 2025 and S.J. 180*, 75th Cong., 1st sess., 1937.
53. "The Scrapmen," *Fortune*, January 1949, 86–91, 134–39.
54. "Steel's Outlook More Promising," *Wall Street Journal*, 1 January 1930; "Big Buyers Raid Scrap Markets," *Wall Street Journal*, 23 May 1931; "The Scrapmen," *Fortune*, January 1949, 86–91, 134–39.
55. *Mansbach Scrap Iron Company v. City of Ashland et al.* Court of Appeals of Kentucky, 235 Ky. 265; 30 S.W.2d 968; 1930 Ky. Lexis-Nexus 338, June 17, 1930, Decided.

CHAPTER 4   ALL US CATS MUST SURELY DO OUR BIT

1. "Scrap Drive Gains Momentum in City," *New York Times*, 22 September 1942.
2. "Scrap Drive Aided by Teen-Age Girls," *New York Times*, 20 February 1943.
3. "President 'Grateful' for Gifts of Scrap," *New York Times*, 25 October 1942.
4. U.S. Department of Commerce Bureau of the Census, *United States Census of Business 1933. Volume I: Summary for the United States* (Washington, DC, 1935), A–4.
5. Ibid., A–20.
6. Studies noting the emphasis Jewish immigrants placed on the education of the second generation include Joel Perlman, "Beyond New York: The Occupations of Russian Jewish Immigrants in Providence, R.I. 1900–1915," *American Jewish History* 72, no. 3 (1983): 369–94; John E. Bodnar, *The Transplanted: A History of Immigrants in Urban America* (Bloomington: Indiana University Press 1985).
7. "Schiavone-Bonomo Marks 50th Year," *Scrap Age* 6, no. 4 (April 1949): 53–55.

8. Bernard Goldman and William Petre, *Navigating the Century: A Personal Account of Alter Company's First Hundred Years* (Chantilly, VA: History Factory, 1998).

9. *Scrap Age Bicentennial Edition*, 51–52.

10. Ibid., 53–55.

11. Cottler's comments come from a roundtable discussion of veteran ISIS members videotaped in January 1986, *Our Heritage: Next Generation II*, available from the Institute of Scrap Recycling Industries (ISRI), Washington, DC.

12. Gutenschwager, "The Scrap Iron and Steel Industry in Metropolitan Chicago," 57–70.

13. Leonard Tanenbaum, *Junk Is Not a Four-Letter Word* (Cleveland: Author, 1993), 163–67.

14. Frank Byrd, 7 December 1938, in Federal Writers' Project, *American Life Histories: Manuscripts from the Federal Writers' Project, 1936–1940*. Library of Congress, Manuscript Division, WPA Federal Writers' Project Collection, Washington, DC.

15. Interview of "The Baron"—"Dutch" Van Bruden by Marian Charles Hatch, 1 November 1938, in Federal Writers' Project, *American Life Histories: Manuscripts from the Federal Writers' Project, 1936–1940*.

16. Tetsuji Okazaki, "The Japanese Iron and Steel Industry, 1929–33, and the Establishment of the Nippon Steel Co.," *Japanese Yearbook on Business History* 4 (1987): 126–51.

17. "The Scrapmen," *Fortune*, January 1949, 86–90, 91, 134.

18. "Roosevelt Moves: Bars Export of Metal to All Except Britain and New World Nations," *New York Times*, 27 September 1940, RG 1631, box 116, folder "Scrap pre-1941," AISI archive, Hagley Museum and Library, Wilmington, DE.

19. Ibid.

20. "Scrap Dearth Held Crippling Japan's Open Hearth Mills, Using Pig Iron in 35 PC Ratio," *New York World Telegram*, 20 August 1941, RG 1631, box 116, folder "Scrap pre-1941," AISI archive, Hagley Museum and Library, Wilmington, DE.

21. Business Press Industrial Scrap Committee, *Primer of Industrial Scrap* (New York, 1942), 4, RG 1631, box 116, folder "Scrap 8," AISI Archive, Hagley Museum and Library, Wilmington, DE.

22. Robert F. Campbell, *The History of Basic Metals: Price Control in World War II* (New York, 1948), 18–20.

23. Ibid., 56–60.

24. Ibid., 60.

25. John A. Hart, *The Beginnings of OPA: Part II. The Price Stabilization Division* (Washington, DC: U.S. Office of Temporary Controls, 1947), 163.

26. Ibid., 164.

27. Ibid., 165.

28. Ibid. Concerns over how well ISIS represented small dealers resurfaced in the postwar period, resulting in the formation of the National Federation of Independent Scrap Yard Dealers in 1955 to represent the smaller yards.

29. Campbell, *The History of Basic Metals*, 74–79.

30. Ibid.

31. Ibid., 77–78.

32. Ibid., 74–79.

33. Institute of Scrap Iron and Steel (ISIS) *Yearbook* (Washington, DC: ISIS, 1942), 17.
34. Campbell, *The History of Basic Metals*, 110–12.
35. Ibid.
36. "OPM Heads Urge a Big Scrap Drive," *New York Times*, 4 September 1941, RG 1631, box 116, folder "Scrap pre-1941," AISI archive, Hagley Museum and Library, Wilmington, DE.
37. Ibid.
38. ISIS, *Scrap Is Fighting Metal*, 5.
39. "Reported War Materials, Inc. Will Soon Open Office In Pittsburgh," *American Metal Market*, 27 August 1942, RG 1631, box 116, folder "Scrap 1941–1949," AISI archive, Hagley Museum and Library, Wilmington, DE.
40. ISIS, *Scrap Is Fighting Metal*.
41. "Iron, Steel Scrap: Foundries and Smelters Need a Lot; Junkyards Will Be Combed for It," *Wall Street Journal*, 15 January 1942, RG 1631, box 116, folder "Scrap 1941–1949," AISI archive, Hagley Museum and Library, Wilmington, DE.
42. "OPA Allows Six Percent Freight Rate Rise to Be Added to Steel Scrap Prices," *Wall Street Journal*, 18 March 1942, RG 1631, box 116, folder "Scrap 1941–1949," AISI archive, Hagley Museum and Library, Wilmington, DE.
43. According to a roundtable interview of veteran dealers, Henderson addressed the 1940 ISIS meeting in Baltimore and accused them all of being thieves; *Our Heritage: Next Generation II*.
44. "Henderson Denies Scrap Ceiling Rise," *New York Times*, 7 October 1941.
45. "Declare Scrap Is Unused," *New York Times*, 28 January 1942.
46. Lexis-Nexis lists 273 separate federal court cases involving the Office of Price Administration, including several cases filed against scrap iron and steel dealers and their customers between 1941 and 1943.
47. "Jones & Laughlin, Allegheny-Ludlum Enjoined On Scrap Price Violations," *American Metal Market*, 20 May 1942, RG 1631, box 116, folder "Scrap 1941–1949," AISI archive, Hagley Museum and Library, Wilmington, DE.
48. "Pittsburgh Steel Company Contends That Prices for Scale and Pit Scrap Are Extra-Regulatory," *American Metal Market*, 18 June 1942, RG 1631, box 116, folder "Scrap 1941–1949," AISI archive, Hagley Museum and Library, Wilmington, DE.
49. "L. B. Block Replies to OPA Charges against Inland," *American Metal Market*, 26 June 1942, RG 1631, box 116, folder "Scrap 1941–1949," AISI archive, Hagley Museum and Library, Wilmington, DE.
50. ISIS, *1945 Yearbook*, 128–29.
51. "Reported War Materials, Inc. Will Soon Open Office in Pittsburgh," *American Metal Market*, 27 August 1942, RG 1631, box 116, folder "Scrap 1941–1949," AISI archive, Hagley Museum and Library, Wilmington, DE.
52. "Inventory Control Is Relaxed by WPB," *New York Times*, 11 December 1943.
53. Campbell, *The History of Basic Metals*, 202.
54. "War Department Reports on Receipts of Battlefield Scrap," *American Metal Market*, 30 July 1943, RG 1631, box 116, folder "Scrap 1941–1949," AISI archive, Hagley Museum and Library, Wilmington, DE.
55. "Army Institutes System for Handling Overseas Scrap," *American Metal Market*, 31 August 1943, RG 1631, box 116, folder "Scrap 1941–1949," AISI archive, Hagley Museum and Library, Wilmington, DE.

56. "Scrap Imported from 30 Countries during Last Year," *American Metal Market*, 2 February 1944, RG 1631, box 116, folder "Scrap 1941–1949," AISI archive, Hagley Museum and Library, Wilmington, DE.

57. Oliver Pollak argues that pressure for second-generation immigrants to find more respectable occupations, as well as changes in the economy, made peddling a marginal activity in Omaha by 1940. Pollak, "The Jewish Peddlers of Omaha," 495.

58. Tanenbaum, *Junk Is Not a Four-Letter Word*, 76.

59. Ibid., 77.

60. Ibid., 78–84.

61. ISIS, *1942 Yearbook*, 100–101.

62. Kenneth Durr and James Lide argue the wartime changes in business practices caused a decline in independent scrap firms. As large corporations increased their recycling of home scrap, Durr and Lide argue, independent collectors and processors were made redundant. Kenneth D. Durr and James H. Lide, "Recycling by Another Name: Business, Efficiency, and the 'New Industrial Philosophy' in World War II," paper presented at American Society for Environmental History Conference, Tucson, AZ, 1999.

CHAPTER 5   SIZE MATTERS

1. "Luria Brothers Initiates Construction for Huge Detroit Scrap Fragmentizer," *American Metal Market*, 15 July 1966, RG 1631, box 116, folder "Luria," AISI archive, Hagley Museum and Library, Wilmington, DE.

2. Lizabeth Cohen, *A Consumers' Republic: The Politics of Mass Consumption in Postwar America* (New York: Alfred A. Knopf, 2003), 121.

3. Adam Ward Rome, *The Bulldozer in the Countryside: Suburban Sprawl and the Rise of American Environmentalism* (New York: Cambridge University Press, 2001), 86.

4. "US to Step Up Auto-Graveyard Scrap Flow; Also Studying Far East Salvage Program," *New York Times*, 4 November 1951, 143; Automobile Manufacturers Association, *Automobile Facts and Figures* (Detroit: Automobile Manufacturers Association, 1971), 24.

5. ISIS, *1947 Yearbook*, 5.

6. Ibid. ISIS observed that federal regulations of the scrap market, already eased with the influx of battlefield scrap in 1944, became less conspicuous to dealers after the war. OPA price ceilings continued into November 1946 but did not generally restrict dealers after the war. Price ceilings resumed during the Korean War, spanning February 1951 to February 1953.

7. American Automobile Manufacturers Association, *Motor Vehicles Facts and Figures*, as quoted in John B. Rae, *The American Automobile Industry* (Boston: G. K. Hall & Co., 1984), 180–82; Automobile Manufacturers Association, *Motor Vehicles Facts and Figures 1998* (Southfield, MI: Automobile Manufacturers Association, 1999), 21.

8. Rae, *The American Automobile Industry*, 62.

9. The earliest reference for the term "automobile graveyard" I have found is in an article concerning roadside cleanup efforts in Westchester County, New York, in 1929. "Map Drive to Save Westchester Beauty," *New York Times*, 27 February 1929, 44.

10. "Automobile Graveyards," *New York Times*, 23 October 1941, 22.

11. "US to Step Up Auto-Graveyard Scrap Flow; Also Studying Far East Salvage Program," *New York Times*, 4 November 1951, 143.

12. W. J. Regan, R. W. James, and T. J. McLeer, "Identification of Opportunities

for Increased Recycling of Ferrous Solid Waste," Report number EPA-SW-45D-72 (Washington, DC: ISIS, 1972), 80–81.

13. Ibid.
14. BOPS accounted for 5 percent of the United States' raw steel output in 1962; 17.4 percent in 1965; almost 50 percent in 1970; and over 60 percent in 1975. Leonard H. Lynn, "Basic-Oxygen Steelmaking Process," in *The Encyclopedia of American Business History and Biography: Iron and Steel in the Twentieth Century*, ed. Bruce Seely (New York City: Facts on File Publications, 1994), 30–32.
15. Survey reprinted in James W. Sawyer Jr., *Automotive Scrap Recycling: Processes, Prices, and Prospects* (Washington, DC: Resources for the Future, 1974), 9.
16. International Institute of Synthetic Rubber Producers, *Synthetic Rubber: The Story of an Industry* (New York: International Institute of Synthetic Rubber Producers, Inc., 1973).
17. "Natural Rubber Supply Too Low," *Waste Trade Journal* 89 (5 August 1950): 49.
18. Thomas E. Leary, "Continuous Casting," *The Encyclopedia of American Business History and Biography: Iron and Steel in the Twentieth Century*, 91–93.
19. "Sagging Scrap: New Steel Technology Helps Cut Waste Metal Sales to Half '56 Level," *Wall Street Journal*, 31 January 1962.
20. The term "waste material dealer"—though used widely—was the source of some consternation among members of the industry during this period. Although the National Association of Waste Material Dealers and the *Waste Trade Journal* retained use of the term in their names, many dealers objected to the identification of their valued, reusable commodities as wastes. NAWMD changed its name to the National Association of Secondary Material Industries (NASMI) in the early 1960s.
21. U.S. Department of Commerce Bureau of the Census, *United States Census of Business 1948. Volume IV: Wholesale Trade—General Statistics Commodity Line Sales Statistics* (Washington, DC: GPO, 1952), 1.27, table 1H, "Merchant Wholesalers—Kind of Business, by Selected Standard Metropolitan Areas: 1948."
22. Ibid., 9, table 2A, "Sales Size—Service Wholesalers—United States, by Kind of Business: 1948."
23. Ibid., 9, table 5A, "Legal Form of Organization—United States, by Type of Operation and Kind of Business."
24. "The Scrapmen," *Fortune*, January 1949, 88.
25. Ibid., 88.
26. Ibid., 90.
27. Ibid., 90, 134.
28. "Scrap and Waste Industry to Spend Millions of Dollars for Equipment," *Waste Trade Journal: Equipment Edition* 85 (July 1948): 7–9.
29. Ibid.
30. Ibid.
31. "Streamlining a Scrap Yard," *Waste Trade Journal: Equipment Edition* 85 (July 1948): 10–13.
32. *Scrap Age Bicentennial Edition*, 56–59.
33. Ibid., 60–64.
34. "Shredders Slowly Reducing Heaps of Junked Cars," *New York Times*, 26 March 1972, 60.

35. R. F. Kuhnlein, "Scrap Preparation and Contamination," *Iron and Steel Engineer* (December 1955): 120–23, RG 1631, box 116, folder "Scrap 1950–1955–3," AISI archive, Hagley Museum and Library, Wilmington, DE.

36. R. F. Kuhnlein, "Scrap Preparation and Contamination," 122.

37. *Scrap Age Bicentennial Edition*, 60–64.

38. Ibid., 65–71. Freight charges increased in the 1970s, further increasing the cost of rail transport. Dealers turned to riverboats, oceangoing vessels, and trucks to minimize transportation costs, and ISIS lobbied the ICC in order to reduce freight charges.

39. ISIS, *1948 Yearbook*, 135–36; idem, *1960 Yearbook*, 52–57.

40. ISIS, *Addresses at the 36th Annual Convention* (Washington, DC: ISIS, 1964), 41–59.

41. "Nathan Trottner Killed by Blast of Pipe in San Antonio Scrap Yard," *Scrap Age* 6, no. 5 (May 1949): 38.

42. Edward Fields, "A Look into the Future by a Small Scrap Dealer," in ISIS, *Addresses at the 19th Annual Convention* (Washington, DC: ISIS, 1947), 26.

43. Ibid., 52.

44. Ibid., 55.

45. Details on the postsale organization of Luria Brothers are available in a study Lukens Steel made of the prospect of investing in the scrap industry. Lukens Steel Company, "The Iron and Steel Scrap Industry as a Diversification Opportunity. December 1957," RG 50, box 2167, folder 13, Hagley Museum and Library, Wilmington, DE.

46. U.S. Department of Commerce Bureau of the Census, *United States Census of Business 1963. Volume IV: Wholesale Trade—General Statistics Commodity Line Sales Statistics* (Washington, DC: GPO, 1966), 7–4, table 1, "Legal Form of Organization—United States, by Type of Business."

47. "Logemann Builds Fully Automatic Scrap Baler for Luria Brothers," *American Metal Market*, 17 August 1965, RG 1631, box 116, AISI archive, Hagley Museum and Library, Wilmington, DE.

48. U.S. Senate Select Committee on Small Business, "Monopoly and Technological Problems in the Scrap Steel Industry," 16 October 1959, 1, RG 1631, box 116, folder "Luria," AISI archive, Hagley Museum and Library, Wilmington, DE.

49. Ibid., 2.

50. Ibid.

51. Ibid.

52. "FTC Is Told Luria's Reliability Is the Reason for Sales," *American Metal Market*, 26 March 1958, RG 1631, box 116, folder "Luria," AISI archive, Hagley Museum and Library, Wilmington, DE.

53. "FTC Aide Rules Steel Scrap Firm, Mills Had Illegal 'Arrangements'," *Wall Street Journal*, 13 April 1961, RG 1631, box 116, folder "Luria," AISI archive, Hagley Museum and Library, Wilmington, DE.

54. Ibid.

55. "FTC Upholds Examiner in Luria Complaint," *American Metal Market*, 3 December 1962, RG 1631, box 116, folder "Luria," AISI archive, Hagley Museum and Library, Wilmington, DE.

56. "Supreme Court Rules on Luria Appeal," *American Metal Market*, 15 October 1968, RG 1631, box 116, folder "Luria," AISI archive, Hagley Museum and Library, Wilmington, DE.

57. "US to Step Up Auto-Graveyard Scrap Flow; Also Studying Far East Salvage Program," *New York Times*, 4 November 1951, 143.

58. American Automobile Manufacturers Association, *Motor Vehicles Facts and Figures*, as quoted in Rae, *The American Automobile Industry*, 180–82.

59. Charles H. Lipsett, *Industrial Wastes and Salvage* (New York: Atlas Publishing Co., 1963), 159.

60. Lady Bird Johnson, quoted in Chester H. Liebs, *Main Street to Miracle Mile* (Boston: Little, Brown and Co., 1985), 65. The First Lady's highway campaign is described in detail in Louis L. Gould, *Lady Bird Johnson: Our Environmental First Lady* (Lawrence: University Press of Kansas, 1999), 90–108.

61. Gould, *Lady Bird Johnson*, 90–108.

62. President Lyndon B. Johnson, "Special Message to the Congress on Conservation and Restoration of Natural Beauty," 8 February 1965. National Archives and Records Administration, Lyndon B. Johnson Library and Museum, accessed at http://www.lbjlib.utexas.edu/johnson/archives.hom/speeches.hom/650208.htm.

63. Joseph A. Califano, *The Triumph and Tragedy of Lyndon Johnson: The White House Years* (College Station: Texas A&M University Press, 2000), 81.

64. "The Problems of Beautification," *Scrap Age* 22, no. 7 (July 1965): 5.

65. "Senate Hearing on 'Beautification' Bill," *Scrap Age* 22, no. 9 (September 1965): 1–2.

66. "It's Time to Wake Up . . . Scrap Is Not Junk!" *Scrap Age* 22, no. 6 (June 1965): 1–3.

67. Ibid.

68. "Beautification—Good or Bad?" *Scrap Age* 22, no. 6 (June 1965): 5.

69. "Highway Beautification Act. It's [*sic*] Effect on Scrap Industry," *Scrap Age* 22, no. 11 (November 1965): 16–17.

70. President Lyndon B. Johnson, "Remarks at the Signing of the Highway Beautification Act of 1965, October 22, 1965." National Archives and Records Administration, Lyndon B. Johnson Library and Museum, Austin, TX, accessed at http://www.lbjlib.utexas.edu/johnson/archives.hom/speeches.hom/651022.asp.

71. Califano, *The Triumph and Tragedy of Lyndon Johnson*, 81.

72. U.S. Senate, *Federal Highway Beautification Assistance Act of 1979: Hearings before the Committee on Transportation of the Committee on Environment and Public Works of the United States Senate* 96th Cong., 1st sess. (Washington, DC: GPO, 1979): 42; Gould, *Lady Bird Johnson*, 103–7.

73. Gould, *Lady Bird Johnson*, 90–108.

74. Ruth Schwartz Cowan, *More Work for Mother: The Ironies of Household Technology from the Open Hearth to the Microwave* (New York: Basic Books, 1983); Hoy, *Chasing Dirt*.

75. Melosi, *The Sanitary City*, 271–73.

76. U.S. Department of Commerce Bureau of the Census, *United States Census of Business 1963*, sect. 7, p. 4, table 1, "Legal Form of Organization—United States, by Type of Business."

77. In the wake of this competition, and frustrated by what some saw as the dominance of ISIS by large brokers, several small yard owners formed the National Federation of Independent Scrap Yard Dealers, a short-lived trade association that members called a way to stay in business and Herman D.

Moskowitz of Schiavone-Bonomo called a cooperative through which small dealers could sell directly to mills and bypass brokerage houses. "Yard Dealers Seek to Improve Quality, Lift Sales through Use of Pilot Plant; Deny They Attempt to By-Pass Brokers," *Scrap Age* 12, no. 6 (June 1955): 80–81.

CHAPTER 6   IT'S NOT EASY BEING GREEN

1. New York City Department of Sanitation Press Release #02–37, 8 July 2002.
2. "NYC Mayor Wants to Dump Recycling," *Associated Press*, 22 April 2002, accessed at http://www.wired.com/news/politics/0,1283,52008,00.html.
3. Ibid.
4. "Is Recycling's Future Behind It?" *New York Times*, 12 March 2002, B–1.
5. Congressional Research Service, "Report for Congress: Bottle Bills and Curbside Recycling: Are They Compatible?" James E. McCarthy, specialist, Environment and Natural Resources Policy Division, 27 January 1993.
6. Melosi, *The Sanitary City*, 411–12.
7. U.S. Environmental Protection Agency, "Characterization of Municipal Solid Waste in the US, 1999 Update" (Washington, DC: U.S. Environmental Protection Agency, 2000), SuDoc: EP1.2: M92/999, record number: ASI 2000 92 14–6.
8. Ibid.
9. Tierney, "Recycling Is Garbage," 29. Several advocates of recycling rebutted Tierney. One of the most comprehensive retorts is Allen Hershkowitz, "In Defense of Recycling," *Social Research* 65, no. 1 (spring 1998): 141–218.
10. U.S. Environmental Protection Agency, "Characterization of Municipal Solid Waste in the US, 1999 Update."
11. Samuel P. Hays, *A History of Environmental Politics since 1945* (Pittsburgh: University of Pittsburgh Press, 2000), 63.
12. ISIS, *Reclamation, Conservation, Beautification* (Washington, DC: ISIS, 1974).
13. Derived from U.S. Environmental Protection Agency, "Characterization of Municipal Solid Waste in the US, 1999 Update."
14. Regan, James, and McLeer, "Identification of Opportunities for Increased Recycling of Ferrous Solid Waste," 177.
15. Ibid., 179.
16. Battelle Memorial Institute, *A Study to Identify Opportunities for Increased Solid Waste Utilization.* Vol. 1. (Washington, DC: U.S. Environmental Protection Agency, 1972), xix, report number EPA-SW-40D.
17. Regan, James, and McLeer, "Identification of Opportunities for Increased Recycling of Ferrous Solid Waste," 162–63. Export quotes established in the early 1970s attempted to curb these numbers, particularly in regards to Japan, but the foreign markets for scrap continued to grow during the 1970s and 1980s.
18. U.S. Bureau of Mines, "Iron and Steel Scrap," *Minerals Yearbook* (Washington, DC: U.S. Bureau of Mines), 1960–70; Regan, James, and McLeer, "Identification of Opportunities for Increased Recycling of Ferrous Solid Waste," 164.
19. Data from U.S. Bureau of Mines, *Annual Reports: 1965.*
20. Texas, Alabama, and other southern states saw an increase in steel production in the 1980s and 1990s, bolstering demand in that region.

21. Automobile Manufacturers Association, *Automobile Facts and Figures*, 24.

22. Regan, James, and McLeer, "Identification of Opportunities for Increased Recycling of Ferrous Solid Waste," 13.

23. U.S. Bureau of the Census, *1982 Census of Wholesale Trade. Geographic Area Series: United States*. (Washington, DC: GPO, 1984), 64, table 5.

24. Regan, James, and McLeer, "Identification of Opportunities for Increased Recycling of Ferrous Solid Waste," 179.

25. Martin T. Katzman, "From Horse Carts to Minimills," *Public Interest* 92 (1988): 124–25.

26. Andrew Hurley, *Environmental Inequalities: Class, Race, and Industrial Pollution in Gary, Indiana, 1945–1980* (Chapel Hill: University of North Carolina Press, 1995).

27. "Little Giants: Minimill Steelmakers, No Longer Very Small, Outperform Big Ones," *Wall Street Journal*, 12 January 1981. As minimills grew larger and more complex, a trend toward an even smaller production facility, the micromill, emerged. With production capacities about a fifth that of a large minimill, micromills were also able to use large quantities of scrap, thus becoming valuable customers to local dealers. "Willy Korf Says Micromill Is the Answer for Small Steelmakers in Glutted Market," *Wall Street Journal*, 7 January 1985.

28. "Miniature Mills Help Clear Land of Rusty Auto Hulks," *Wall Street Journal*, 26 October 1981.

29. Battelle Memorial Institute, *A Study to Identify Opportunities for Increased Solid Waste Utilization*, ix–x.

30. "Shredders Slowly Reducing Heaps of Junked Cars," *New York Times*, 26 March 1972, 60.

31. Battelle Memorial Institute, *A Study to Identify Opportunities for Increased Solid Waste Utilization*, ix.

32. Ibid., xx–xxi.

33. Melosi, *The Sanitary City*, 350.

34. Battelle Memorial Institute, *A Study to Identify Opportunities for Increased Solid Waste Utilization*, xx–xxi.

35. Arthur J. Warner, Charles H. Parker, and Bernard Baum, *Solid Waste Management of Plastics* (Washington, DC: Manufacturing Chemicals Association, December 1970), 1–2.

36. Rachel Carson, *Silent Spring* (Greenwich, CT: Fawcett Publications, 1962).

37. Joel Tarr argues that these new values reflect the findings of the discipline of environmental health, with its focus on chronic, degenerative diseases. In this regard, Tarr argues, we have returned to the original thrust of the nineteenth-century sanitary movement toward cleansing the urban environment in order to ensure freedom from epidemics. Tarr, *The Search for the Ultimate Sink*, 351.

38. Hays, *A History of Environmental Politics since 1945*, 55.

39. Robert Gottlieb, *Forcing the Spring: The Transformation of the American Environmental Movement* (Washington, DC: GPO, 1993): 81–86, 132.

40. Melosi, *The Sanitary City*, 350–54; Tarr, *The Search for the Ultimate Sink*, 26.

41. Gottlieb, *Forcing the Spring*, 129; Regan, James, and McLeer, "Identification of Opportunities for Increased Recycling of Ferrous Solid Waste," 47–48; Jack Lewis, "The Birth of EPA," *EPA Journal* (November 1985): 9, accessed at http://www.epa.gov/history/topics/epa/15c.htm.

42. Angela S. Wilkes, Irene Kiefer, and Barbara Levine, *Everybody's Problem:*

Hazurdous Waste (Washington, DC: U.S. Environmental Protection Agency, Office of Water and Waste Management, 1980).

43. Resource Conservation and Recovery Act 42 USC s/s 6901 et seq. (1976). US EPA, "Summary of Resource Conservation and Recovery Act," accessed at http://www.epa.gov/region5/defs/html/rcra.htm.

44. William J. Becker, "OSHA History, Purposes, and Activities," Fact Sheet AE–114, University of Florida, Florida Cooperative Extension Service, August 1992.

45. Joseph S. Schapiro, "Accidents—Especially Those That Must Be Avoided—Can Be Avoided," *Scrap Age* 33, no. 7 (July 1976): 39–41, 46.

46. Battelle Memorial Institute, *A Study to Identify Opportunities for Increased Solid Waste Utilization*, 7.

47. Ibid.

48. Tarr, *The Search for the Ultimate Sink*, 349; Council on Environmental Quality (CEQ), *Environmental Quality—1979* (Washington, DC: GPO, 1979), 174–88; CEQ, *Environmental Quality—1980* (Washington, DC: GPO, 1980), 214–22.

49. Katzman, "From Horse Carts to Minimills," 131–32.

50. Battelle Memorial Institute, *A Study to Identify Opportunities for Increased Solid Waste Utilization*, 11–12.

51. "Recyclers Sue ICC," *Chemical Week*, 6 June 1979, 36.

52. "Exclusive Interview Harlan Carroll President NARI 1980," *Scrap Age* 37, no. 4 (April 1980): 30–45.

53. A. V. Bridgewater and C. J. Mumford, *Waste Recycling and Pollution Control Handbook* (New York: Van Nostrand Reinhold, 1979), 430–31.

54. Discussions of the hazardous materials contained in automobiles are found in Thomas E. Graedel and Braden R. Allenby, *Industrial Ecology and the Automobile* (Upper Saddle River, NJ: Prentice Hall, 1998); Claire B. Kendrick, "Steel Scrap: Meeting the Metallic Demand of the Steel Industry" (Ph.D. diss, Pennsylvania State University, 1991), 41.

55. "Soil Cleanup to Begin at Site of Bankrupt Scrap Company in Minneapolis," *Minneapolis Star Tribune*, 5 April 1988.

56. "EPA Adds Spokane Junkyard to List of Toxic Sites: Owner, Dumpers Will Be Charges for Cleanup," *Spokane Spokesman-Review*, 2 June 1994.

57. "A Scrap Over the Superfund Law," *Nation's Business* 80, no. 7 (July 1992): 59.

58. Ibid.

59. Superfund Recycling Equity, 42 USC § 9627.SEC. 6001, accessed at http://es.epa.gov/oeca/osre/recycle.html.

60. Katzman, "From Horse Carts to Minimills," 121–25. On Superfund, see John A. Hird, *Superfund: The Political Economy of Environmental Risk* (Baltimore: Johns Hopkins University Press, 1994).

61. Interview of William Breman, president of Breman Steel by Carl Zimring, 27 August 1999. Berman cites costs of regulation and new technology as factors in the decisions of many scrap dealers (including him) to sell their companies over the past quarter century. Dealer complaints from roundtable interview *Our Heritage: Next Generation II*.

62. Battelle Memorial Institute, *A Study to Identify Opportunities for Increased Solid Waste Utilization*, 2.

63. Strasser, *Waste and Want*, 284.

64. Firms attempted to modernize more than their processing technology and transportation. Yard security for many firms was a low-tech operation, yet

even this aspect of the industry evolved over the 1970s. Wrecking yards featured valuable materials often sitting outside in plain view. Yard owners secured their wares with barbed wire fences and guard dogs. The dogs, usually German shepherds or pit bulls, were selected to be as intimidating as possible. This aspect of the industry, one of the more recognizable to the public for decades, grew in sophistication by 1980; instead of purchasing and training their own dogs, yard owners could lease trained junkyard dogs for night shifts from agencies. Katzman, "From Horse Carts to Minimills," 127.

65. "Exclusive Interview Morton Plant President ISIS 1980," *Scrap Age* 37, no. 2 (February 1980): 67, 69, 71, 73, 75, 77.

66. Much as was the case when Luria Brothers was sold, SHV Holdings retained David J. Joseph's managers, opting to name David J. Joseph Jr. chairman of its new subsidiary in 1978. "Commerce and Industry," *Wall Street Journal*, 6 April 1978; "Heap of Trouble Hits Scrap Metal Business," *Pittsburgh Post-Gazette*, 30 October 1998.

67. Data from U.S. Bureau of Mines *Annual Reports*, 1965, 1980. Available from the US Geological Survey website, at http://minerals.usgs.gov/minerals/pubs/commodity/iron_&_steel_scrap/stat/tbl1.txt.

68. One dealer noted that interest in hazardous material abatement allowed him to enter asbestos removal, where "the regulations are unbelievable, but what's good is you then charge your customers for upkeep for a hell of a price!" Roundtable discussion of veteran ISIS members in *Our Heritage: Next Generation II.*

69. Battelle Memorial Institute, *A Study to Identify Opportunities for Increased Solid Waste Utilization*, x.

70. Ibid., xxii.

71. Ibid., xxiii.

72. Ibid.

73. Data from U.S. Bureau of Mines, *Annual Reports*, 1965, 1980. Quoted in the US Geological Survey website, at http://minerals.usgs.gov/minerals/pubs/commodity/iron_&_steel_scrap/stat/tbl1.txt; Institute of Scrap Recycling Industries, "Mission Statement" (Washington, DC: Institute of Scrap Recycling Industries, 2000).

74. Katzman, "From Horse Carts to Minimills," 125.

75. Ibid.; "Mining the Scrap Heap for Treasure," *Smithsonian Magazine*, May 1997.

76. "City to Resume Recycling of Plastics," *New York Times*, 14 January 2003, B–1.

77. "Gold in Them Thar Tin Cans? Recycler Sees Money to Be Made from City's Containers," *New York Times*, 11 January 2003, B–1.

CONCLUSION

1. Rathje and Murphy, *Rubbish!* 204.

2. Most scrap processing in Chicago continued to be done in largely African American neighborhoods on the South Side and West Side, continuing racial segregation of environmental inequalities that had persisted since the early twentieth century. For an excellent case study of how these inequalities replicated in commercial and grassroots recycling programs, see David Naguib Pellow, *Garbage Wars.*

3. Perry, *Collecting Garbage.*

4. *The Sopranos* (television series), Home Box Office, 1999. Actual organized-crime cases involving New Jersey waste firms are detailed in unreliable fashion in Alan A. Block and Frank R. Scarpitti, *Poisoning for Profit: The Mafia and Toxic Waste in America* (New York: William Morrow & Co., 1985). Critics remarked on the derogatory nature of the junk dealer in *Star Wars: Episode One* at the time of the film's release, noting it conformed to nineteenth-century stereotypes of hook-nosed, shifty Jews. "The Merchant of Menace: Racial Stereotypes in a Galaxy Far, Far Away?" Bruce Gottlieb, *Slate*, 26 May 1999, at http://slate.msn.com/?id=29394.

5. Donald Reid, *Paris Sewers and Sewermen: Realities and Representations* (Cambridge, MA: Harvard University Press, 1991).

6. William McDonough and Michael Braungart, *Cradle to Cradle: Remaking the Way We Make Things* (New York: North Point Press, 2002); Harvey Molotch, *Where Stuff Comes From: How Toasters, Toilets, Cars, Computers, and Many Other Things Come to Be as They Are* (New York and London: Routledge, 2003).

7. T. J. Jackson Lears, *Fables of Abundance: A Cultural History of Advertising in America* (New York: Basic Books, 1994).

Bibliography

PRIMARY SOURCES

Alphabetical Trade Catalog Collection. Smithsonian Institute Libraries, Washington, DC.

American Iron and Steel Archive. Hagley Museum and Library, Wilmington, DE.

Becker, William J. "OSHA History, Purposes, and Activities." Fact Sheet AE-114. University of Florida, Florida Cooperative Extension Service. August 1992.

Congressional Research Service. *Report for Congress: Bottle Bills and Curbside Recycling: Are They Compatible?* Written by James E. McCarthy, Specialist, Environment and Natural Resources Policy Division. Washington, DC: Congressional Research Service, 1993.

Council on Environmental Quality. *Environmental Quality—1979.* Washington, DC: GPO, 1979.

———. *Environmental Quality—1980.* Washington, DC: GPO, 1980.

Federal Writers' Project. *American Life Histories: Manuscripts from the Federal Writers' Project, 1936–1940.* Library of Congress, Manuscript Division, WPA Federal Writers' Project Collection, Washington, DC.

Francis Bannerman Son, Inc., Archive. Hagley Museum and Library, Wilmington, DE.

George P. Whitaker Co. Papers. MS 1730.1. Maryland Historical Society Library, Baltimore.

Heinz Collection of Machine Tool and Metal Working Trade Literature. Smithsonian Institute Libraries, Washington, DC.

Institute of Scrap Iron and Steel (ISIS). Addresses at the National Convention. Institute of Scrap Recycling Industries Headquarters, Washington, DC.

———. Yearbook. Institute of Scrap Recycling Industries Headquarters, Washington, DC.

Lukens Steel Company Archive. Hagley Museum and Library, Wilmington, DE.

National Council of Jewish Women, Pittsburgh Section, Oral History Project #1. Archives of Industrial Society, University of Pittsburgh, Pittsburgh, PA.

Noyes, Morillo. Manuscripts. Baker Library, Harvard Business School, Cambridge, MA.

U.S. Bureau of Mines. *Annual Reports*. Washington, DC: GPO, 1965, 1980.

———. "Iron and Steel Scrap." *Minerals Yearbook*. Vol. 1. Washington, DC: GPO, 1960–70.

U.S. Congress. Senate. Committee on Military Affairs. 1937. *Hearings before a Subcommittee of the Committee on Military Affairs on S. 2025 and S.J. 180.* 75th Cong., 1st sess. CIS No.: 75 5549-2 SuDoc: Y4.M59/2: Ser 16.

———. Senate. *Federal Highway Beautification Assistance Act of 1979: Hearings before the Committee on Transportation of the Committee on Environment and Public Works of the United States Senate.* 96th Cong., 1st sess. Serial No. 96-H25 SuDoc: Y4.P96/10:96-H25.

———. Senate. Select Committee on Small Business. 1959. *Monopoly and Technological Problems in the Scrap Steel Industry.* 86th Cong., 1st sess. CIS-No.: 86 51359-6 SuDoc: Y4.Sm1/2: Ser 1.

U.S. Department of Commerce. Bureau of the Census. *Elimination of Waste: Classification of Iron and Steel Scrap.* Washington, DC: GPO, 1926.

———. *United States Census of Business 1933. Volume I: Summary for the United States.* Washington, DC: GPO, 1935.

———. *United States Census of Business 1948. Volume IV: Wholesale Trade—General Statistics Commodity Line Sales Statistics.* Washington, DC: GPO, 1952.

———. *United States Census of Business 1963. Volume IV: Wholesale Trade—General Statistics Commodity Line Sales Statistics.* Washington, DC: GPO, 1966.

U.S. Environmental Protection Agency. *A Study to Identify Opportunities for Increased Solid Waste Utilization.* Vol. 1. Report no. EPA-SW-40D. Washington, DC, 1972.

———. "Summary of Resource Conservation and Recovery Act." Resource Conservation and Recovery Act 42 U.S.C. s/s 6901 et seq. (1976). http://www.epa.gov/region5/defs/html/rcra.htm. Accessed 8 January 2000.

———. Superfund Recycling Equity. 42 U.S.C. § 9627.SEC. 6001. http://es.epa.gov/oeca/osre/recycle.html. Accessed 4 March 2002.

U.S. Office of Temporary Controls. *The Beginnings of OPA: Part II. The Price Stabilization Division.* Written by John A. Hart. Washington, DC: GPO, 1947.

SECONDARY SOURCES

Allen, David T., and Nasrin Behmanesh. "Wastes as Raw Materials." In *The Greening of Industrial Ecosystems,* ed. Braden R. Allenby and Deanna J. Richards, 69–89. Washington, DC: National Academy Press, 1994.

American Iron and Steel Association. *Annual Statistical Report of the American Iron and Steel Association.* Philadelphia: American Iron and Steel Association, 1888.

———. *Report of the Secretary of the American Iron and Steel Association.* Philadelphia: American Iron and Steel Association, 1872.

Automobile Manufacturers Association. *Automobile Facts and Figures.* Detroit: Automobile Manufacturers Association, 1971.

———. *Motor Vehicles Facts and Figures.* Southfield, MI: Automobile Manufacturers Association, 1999.

Ball, John M. *Reclaimed Rubber: The Story of an American Raw Material.* New York: Rubber Reclaimers Association, 1947.

Barringer, Edwin C. *The Story of Scrap.* Washington, DC: Institute of Scrap Iron & Steel, 1954.

Battelle Memorial Institute. *A Study to Identify Opportunities for Increased Solid Waste Utilization.* Report Number EPA-SW-40D: xix. Washington, DC: U.S. Environmental Protection Agency, 1972.

Becker, William J. "OSHA History, Purposes, and Activities." Fact Sheet AE-114, University of Florida, Florida Cooperative Extension Service, August 1992.

Bell, Thomas. *Out of This Furnace.* Boston: Little, Brown and Co., 1941.

Benson, Susan Porter. *Counter Cultures: Saleswomen, Managers, and Customers in American Department Stores, 1890–1940.* Urbana: University of Illinois Press, 1986.

Blackford, Mansel G., and Austin K. Kerr. *Business Enterprise in American History.* Boston: Houghton Mifflin, 1986.

Block, Alan A., and Frank R. Scarpitti. *Poisoning for Profit: The Mafia and Toxic Waste in America.* New York: W. Morrow, 1985.

Blumin, Stuart M. *The Emergence of the Middle Class: Social Experience in the American City, 1760–1900.* New York: Cambridge University Press, 1989.

Bodnar, John. *The Transplanted: A History of Immigrants in Urban America.* Bloomington: Indiana University Press, 1985.

Bodnar, John, Roger Simon, and Michael P. Weber. *Lives of Their Own: Blacks, Italians and Poles in Pittsburgh, 1900–1960.* Urbana: University of Illinois Press, 1982.

Bonacich, Edna. "Middleman Minorities and Advanced Capitalism." *Ethnic Groups* 2 (1980): 211–19.

Bridgewater, A. V., and C. J. Mumford. *Waste Recycling and Pollution Control Handbook.* New York: Van Nostrand Reinhold, 1979.

Califano, Joseph A. *The Triumph and Tragedy of Lyndon Johnson: The White House Years.* College Station: Texas A&M University Press, 2000.

Campbell, Robert F. *The History of Basic Metals: Price Control in World War II.* New York: Columbia University Press, 1948.

Carless, Jennifer. *Taking out the Trash: A No-Nonsense Guide to Recycling.* Washington, DC: Island Press, 1992.

Carson, Rachel. *Silent Spring.* Greenwich, CT: Fawcett Publications, 1962.

Chandler, Alfred D., Jr. *The Visible Hand: The Managerial Revolution in American Business.* Cambridge, MA: Belknap Press, 1977.

Chase, Stuart. *The Tragedy of Waste.* New York: Macmillan Co., 1927.

Clapperton, Robert H. *The Paper-making Machine.* New York: Pergamon Press, 1967.

Cohen, Lizabeth. *A Consumers' Republic: The Politics of Mass Consumption in Postwar America.* New York: Alfred A. Knopf, 2003.

Cowan, Ruth Schwartz. *More Work for Mother: The Ironies of Household Technology from the Open Hearth to the Microwave.* New York: Basic Books, 1983.

Cronon, William. *Nature's Metropolis: Chicago and the Great West.* New York: W. W. Norton, 1991.

Daniels, Roger. *Coming to America: A History of Immigration and Ethnicity in American Life.* New York: HarperCollins, 1990.

Desrouchers, Pierre. "Market Processes and the Closing of 'Industrial Loops':

A Historical Reappraisal." *Journal of Industrial Ecology* 4 (winter 2000): 29–43.

Diamant, Lincoln. "The Great Chain Hoax." *Hudson Valley Regional Review* 7.1 (March 1990): 44–57.

Diner, Hasia R. *A Time for Gathering: Second Migration, 1820–1880.* Baltimore: Johns Hopkins University Press, 1992.

Douglas, Mary. *Purity and Danger: An Analysis of Concepts of Pollution and Taboo.* New York: Frederick A. Praeger Publishers, 1966.

Durr, Kenneth D., and James H. Lide. "A 'New Industrial Philosophy'?: World War II and the Roots of Corporate Recycling." Paper presented at American Society for Environmental History Conference, Tucson, AZ, 1999.

Fleming, George T. *History of Pittsburgh and Environs, from Prehistoric Days to the Beginning of the American Revolution.* New York: American Historical Society, 1922.

Fontaine, Laurence. *History of Pedlars in Europe.* Trans. Vicki Whittaker. Durham, NC: Duke University Press, 1996.

Forbes, Esther. *Paul Revere and the World He Lived In.* New York: Houghton Mifflin, 1969.

Fukuzawa, Yukichi. *The Autobiography of Yukichi Fukuzawa.* Trans. Eiichi Kiyooka. New York: Columbia University Press, 1966.

Galambos, Louis. *Competition and Cooperation: The Emergence of a National Trade Association.* Baltimore: Johns Hopkins Press, 1966.

Goldman, Bernard, and William Petre. *Navigating the Century: A Personal Account of Alter Company's First Hundred Years.* Chantilly, VA: History Factory, 1998.

Gorman, Hugh S. *Redefining Efficiency: Pollution Concerns, Regulatory Mechanisms, and Technological Change in the U.S. Petroleum Industry.* Akron: University of Akron Press, 2001.

Gottlieb, Bruce. "The Merchant of Menace: Racial Stereotypes in a Galaxy Far, Far Away?" *Slate* (26 May 1999). http://slate.msn.com/?id=29394. Accessed 14 September 2000.

Gottlieb, Robert. *Forcing the Spring: The Transformation of the American Environmental Movement.* Washington, DC: Island Press, 1993.

Gould, Louis L. *Lady Bird Johnson: Our Environmental First Lady.* Lawrence: University Press of Kansas, 1999.

Graedel, Thomas E., and Braden R. Allenby. *Industrial Ecology and the Automobile.* Upper Saddle River, NJ: Prentice Hall, 1998.

Greenstein, Lillian R. "The Peddlers of Bay City." *Michigan Jewish History* 25 nos. 1–2 (1985): 10–17.

Grigg, Harry H., and George E. Haynes. *Junk Dealing and Juvenile Delinquency.* Text by Albert E. Webster. Chicago: Juvenile Protective Association, c. 1919.

Gutenschwager, Gerald A. "The Scrap Iron and Steel Industry in Metropolitan Chicago." Ph.D. diss., University of Chicago, 1957.

Halttunen, Karen. *Confidence Men and Painted Women: A Study of Middle-Class Culture in America, 1830–1870.* New Haven: Yale University Press, 1982.

Hart, John A. *The Beginnings of OPA: Part II. The Price Stabilization Division.* Washington, DC: U.S. Office of Temporary Controls, 1947.

Hays, Samuel P. *Conservation and the Gospel of Efficiency.* Cambridge, MA: Harvard University Press, 1959.

———. *A History of Environmental Politics since 1945.* Pittsburgh: University of Pittsburgh Press, 2000.

Hershkowitz, Allen. "In Defense of Recycling." *Social Research* 65 (spring 1998): 141–218.

Hird, John A. *Superfund: The Political Economy of Environmental Risk*. Baltimore: Johns Hopkins University Press, 1994.

Hogan, William T. *Economic History of the Iron and Steel Industry in the United States*. Lexington, MA: Heath, 1971.

Hoy, Suellen M. *Chasing Dirt: The American Pursuit of Cleanliness*. New York: Oxford University Press, 1995.

Hoy, Suellen M., and Michael C. Robinson. *Recovering the Past: A Handbook of Community Recycling Programs, 1890–1945*. Chicago: Public Works Historical Society, 1979.

Hurley, Andrew. *Environmental Inequalities: Class, Race, and Industrial Pollution in Gary, Indiana, 1945–1980*. Chapel Hill: University of North Carolina Press, 1995.

Hyde, Charles K. *Copper for America: The United States Copper Industry from Colonial Times to the 1990s*. Tucson: University of Arizona Press, 1998.

Institute of Scrap Recycling Industries. *The Original Recyclers*. Videotape, June, 1980. Washington, DC: Institute of Scrap Recycling Industries.

———. *Our Heritage: Next Generation II*. Videotape, January, 1986. Washington, DC: Institute of Scrap Recycling Industries.

International Institute of Synthetic Rubber Producers. *Synthetic Rubber: The Story of an Industry*. New York: International Institute of Synthetic Rubber Producers, 1973.

Joyce, Joseph A., and Howard C. Joyce. *Treatise on the Law Governing Nuisances: With Particular Reference to Its Application to Modern Conditions and Covering the Entire Law Relating to Public and Private Nuisances, Including Statutory and Municipal Powers and Remedies, Legal and Equitable*. Albany: M. Bender & Co., 1906.

Kasson, John F. *Rudeness and Civility: Manners in Nineteenth-Century Urban America*. New York: Hill and Wang, 1990.

Katzman, Martin T. "From Horse Carts to Minimills." *Public Interest* 92 (1988): 124–25.

Kendrick, Claire B. "Steel Scrap: Meeting the Metallic Demand of the Steel Industry." Ph.D. diss, Pennsylvania State University, 1991.

Kimball, Debi. *Recycling in America: A Reference Handbook*. Santa Barbara, CA: ABC-CLIO, 1992.

Kouwenhoven, John A. *The Beer Can by the Highway: Essays on What's American about America*. New York: Doubleday & Co., 1961. Reprint, Baltimore: Johns Hopkins University Press, 1988.

Kraut, Alan M. "The Butcher, The Baker, The Pushcart Peddler: Jewish Foodways and Entrepreneurial Opportunity in the East European Immigrant Community, 1880–1940." *Journal of American Culture* 6 (1983): 71–83.

Lathrop, William Gilbert. *The Brass Industry in the United States: A Study of the Origin and Development of the Brass Industry in the Naugatuck Valley and Its Subsequent Extension over the Nation*. Mount Caramel, CT: W. G. Lathrop, 1926.

Laws and Ordinances Ordained and Established by the Mayor, Aldermen, and Commonalty of the City of New York in Common Council Convened. New York: City of New York, 1817.

Lears, T. J. Jackson. *Fables of Abundance: A Cultural History of Advertising in America*. New York: Basic Books, 1994.

Leary, Thomas E. "Continuous Casting." In *The Encyclopedia of American Busi-*

ness History and Biography: Iron and Steel in the Twentieth Century, ed. Bruce Seely, 91–93. New York: Facts on File Publications, 1994.

Levinson, Aaron P. *If Only Right Now Could Be Forever.* Hillsboro, OR: Media Weavers, 1987.

Lewis, Jack. "The Birth of EPA." *EPA Journal* (November 1985). http://www.epa.gov/history/topics/epa/15c.htm. Accessed 23 May 2001.

Licht, Walter. *Industrializing America: The Nineteenth Century.* Baltimore: Johns Hopkins University Press, 1995.

Liebs, Chester H. *Main Street to Miracle Mile.* Boston: Little, Brown and Co., 1985.

Light, Ivan, and Steven J. Gold. *Ethnic Economies.* San Diego: Academic Press, 2000.

Lipsett, Charles H. *Industrial Wastes and Salvage.* New York: Atlas Publishing Co., 1963.

———. *100 Years of Recycling History: From Yankee Tincart Peddlers to Wall Street Scrap Giants.* New York: Atlas Publishing Co., 1974.

Lukens Steel Company, "The Iron and Steel Scrap Industry as a Diversification Opportunity. December 1957." Hagley Museum and Library, Wilmington, DE. Group 50, Box 2167, Folder 13.

Lynn, Leonard H. "Basic-Oxygen Steelmaking Process." In *The Encyclopedia of American Business History and Biography: Iron and Steel in the Twentieth Century,* ed. Bruce Seely, 30–32. New York City: Facts on File Publications, 1994.

MacLeish, Archibald. *Jews in America.* New York: Random House, 1936.

Maher, John William. "Retrieving the Obsolete: Formation of the American Scrap Steel Industry, 1870–1933." Ph.D. diss., University of Maryland, 1999.

Marchand, Roland. *Advertising the American Dream: Making Way For Modernity, 1920–1940.* Berkeley: University of California Press, 1985.

McCarthy, Thomas Martin. "The Road to Respect: Americans, Automobiles, and the Environment." Ph.D. diss., Yale University, 2001.

McDonough, William, and Michael Braungart. *Cradle to Cradle: Remaking the Way We Make Things.* New York: North Point Press, 2002.

McGaw, Judith A. *Most Wonderful Machine: Mechanization and Social Change in Berkshire Paper Making, 1801–1885.* Princeton, NJ: Princeton University Press, 1987.

Melosi, Martin V. *Garbage in the Cities: Refuse, Reform, and the Environment, 1880–1980.* College Station: Texas A&M University Press, 1981.

———. *The Sanitary City: Urban Infrastructure in America from Colonial Times to the Present.* Baltimore: Johns Hopkins University Press, 2000.

Molotch, Harvey. *Where Stuff Comes From: How Toasters, Toilets, Cars, Computers and Many Other Things Come to Be as They Are.* New York and London: Routledge, 2003.

Naggar, Betty. *Jewish Pedlars and Hawkers, 1740–1940.* Camberley, England: Porphyrogenitus, 1992.

———. "Old-Clothes Men: 18th and 19th Centuries." *Jewish Historical Studies* 31 (1988–90): 171–91.

Nasaw, David. *Children of the City: At Work and at Play.* Garden City, NY: Anchor Press/Doubleday, 1985.

National Association of Waste Material Dealers. *Fifteenth Anniversary Blue Book.* New York: National Association of Waste Material Dealers, 1928.

———. *Twenty-Fifth Anniversary Blue Book.* New York: National Association of Waste Material Dealers, 1938.

Nelson, Daniel. *Frederick W. Taylor and the Rise of Scientific Management.* Madison: University of Wisconsin Press, 1980.

Norris, Frank. *McTeague.* New York: Doubleday and McClure, 1899.

Okazaki, Tetsuji. "The Japanese Iron and Steel Industry, 1929–33, and the Establishment of the Nippon Steel Co." *Japanese Yearbook on Business History* 4 (1987): 126–51.

Pellow, David Naguib. *Garbage Wars: The Struggle for Environmental Justice in Chicago.* Cambridge, MA: MIT Press, 2002.

Perlman, Joel. "Beyond New York: The Occupations of Russian Jewish Immigrants in Providence, R.I. 1900–1915." *American Jewish History* 72.3 (1983): 369–94.

Perry, Stuart E. *Collecting Garbage: Dirty Work, Clean Jobs, Proud People.* New Brunswick, NJ: Transaction Publishers, 1998.

Pollak, Oliver B. "The Jewish Peddlers of Omaha." *Nebraska History* 63 (1982): 474–501.

Prain, Sir Ronald. *Copper: The Anatomy of an Industry.* London: Mining Journal Books, 1975.

Preston, Samuel H., Ewbank, Douglas, and Hereward, Mark. "Child Mortality Differences by Ethnicity and Race in the United States: 1900–1910." In *After Ellis Island: Newcomers and Natives in the 1910 Census,* ed. Susan Cotts Watkins, 35–82. New York: Russell Sage Foundation, 1994.

Rae, John B. *The American Automobile Industry.* Boston: G. K. Hall & Co., 1984.

Rathje, William, and Cullen Murphy. *Rubbish! The Archaeology of Garbage.* New York: HarperCollins, 1992.

Regan, W. J., R. W. James, and T. J. McLeer. "Identification of Opportunities for Increased Recycling of Ferrous Solid Waste." Report number EPA-SW–45D-72. Washington, DC: Institute of Scrap Iron and Steel, 1972.

Reid, Donald. *Paris Sewers and Sewermen: Realities and Representations.* Cambridge, MA: Harvard University Press, 1991.

Richler, Mordecai. *The Apprenticeship of Duddy Kravitz.* Middlesex: Penguin Books, 1964.

Richmond, Craig McMillan. "Simulating Differences in Ferrous Scrap Prices Over Geographic Space Using the Logistic Model of Choice for Differentiated Products." Ph.D. diss., University of Pittsburgh, 1997.

Rome, Adam Ward. *The Bulldozer in the Countryside: Suburban Sprawl and the Rise of American Environmentalism.* New York: Cambridge University Press, 2001.

Rosen, Christine. "Differing Perceptions of the Value of Pollution Abatement across Time and Place: Balancing Doctrine in Pollution Nuisance Law, 1840–1906." *Law and History Review* 2 (fall 1993): 303–81.

———. "Industrial Ecology and the Greening of Business History." *Business and Economic History* 26 (fall 1997): 123–37.

Rosen, George. *A History of Public Health.* New York: MD Publications, 1958.

Rothman, Hal K. *Saving the Planet: The American Response to the Environment in the Twentieth Century.* Chicago: Ivan R. Dee, 2000.

Sawyer, James W., Jr. *Automotive Scrap Recycling: Processes, Prices, and Prospects.* Washington, DC: Resources for the Future, 1974.

Schmier, Louis. "'For Him the "Schwartzers" Couldn't Do Enough': A Jewish Peddler and His Black Customers Look at Each Other." *American Jewish History* 73, no. 1 (1983): 39–55.

Schneider, Eric. *In the Web of Class: Delinquents and Reformers in Boston, 1810s–1930s.* New York: New York University Press, 1992.

Scranton, Philip B. *Endless Novelty: Specialty Production and American Industrialization, 1865–1925.* Princeton, NJ: Princeton University Press, 1997.

Scrap Age Bicentennial Edition. Niles, IL: Three Sons Publishing Co., 1977.

Shammas, Carole. *The Pre-Industrial Consumer in England and America.* New York: Clarendon Press, 1990.

Sinclair, Upton. *The Jungle.* New York: Doubleday, Page & Co., 1906.

Sorin, Gerald. *A Time for Building: The Third Migration, 1880–1920.* Baltimore: Johns Hopkins University Press, 1992.

Stansell, Christine. *City of Women: Sex and Class in New York, 1789–1860.* Urbana: University of Illinois Press, 1987.

Strasser, Susan. *Waste and Want: A Social History of Trash.* New York: Metropolitan Books, 1999.

Tanenbaum, Leonard. *Junk Is Not a Four-Letter Word.* Cleveland: Author, 1993.

Tarr, Joel A. *The Search for the Ultimate Sink: Urban Pollution in Historical Perspective.* Akron, OH: University of Akron Press, 1996.

Thompson, Michael. *Rubbish Theory: The Creation and Destruction of Value.* Oxford: Oxford University Press, 1979.

Thurston, George H. *Pittsburgh and Allegheny in the Centennial Year.* Pittsburgh: A. A. Anderson & Son, 1876.

Tierney, John. "Recycling Is Garbage." *New York Times Magazine*, 30 June 1996, 24–29, 44, 48–49, 53.

Tobey, Ronald C. *Technology as Freedom: The New Deal and the Electrical Modernization of the American Home.* Berkeley: University of California Press, 1996.

Wade, Richard C. *The Urban Frontier: The Rise of Western Cities, 1790–1830.* Cambridge, MA: Harvard University Press, 1959.

Waldinger, Roger D. *Through the Eye of the Needle: Immigrants and Enterprise in New York's Garment Trades.* New York: New York University Press, 1986.

Waldinger, Roger D., Howard Aldrich, Robin Ward, with the collaboration of Jochen Blaschke, Jeremy Boissevain, William D. Bradford, Gavin Chen, Hanneke Grotenbreg, Isaak Joseph, Hermann Korte, Ivan Light, David McEnvoy, Mirjana Morokvasic, Annie Phizacklea, Marlene Sway, Phina Werbner, and Peter Wilson. *Ethnic Entrepreneurs: Immigrant Business in Industrial Societies.* Newbury Park, CA: Sage Publications, 1990.

Ward, David. *Cities and Immigrants: A Geography of Change in Nineteenth Century America.* New York: Oxford University Press, 1971.

Warner, Arthur J., Charles H. Parker, and Bernard Baum. *Solid Waste Management of Plastics.* Washington, DC: Manufacturing Chemicals Association, 1970.

Watkins, Susan Cotts. "Background: About the 1910 Census." In *After Ellis Island: Newcomers and Natives in the 1910 Census*, ed. Susan Cotts Watkins, 11–34. New York: Russell Sage Foundation, 1994.

Weinberg, Adam S., David N. Pellow, and Allan Schnaiberg. *Urban Recycling and the Search for Sustainable Community Development.* Princeton, NJ: Princeton University Press.

Wilkes, Angela S., Irene Kiefer, and Barbara Levine. *Everybody's Problem: Hazardous Waste.* Washington, DC: U.S. Environmental Protection Agency, Office of Water and Waste Management, 1980.

Wilson, Harold S. *Confederate Industry: Manufacturers and Quartermasters in the Civil War.* Jackson: University Press of Mississippi, 2002.

Wolf, Howard. *The Story of Scrap Rubber.* Akron, OH: A. Schulman, 1943.

Wood, Horace Gay. *A Practical Treatise on the Law of Nuisances in Their Various Forms.* Albany: John D. Parsons, 1875.

Zenner, Walter P. "Introduction: Symposium on Economics and Ethnicity: The Case of the Middleman Minorities." *Ethnic Groups* 2 (1980): 185–87.

Zimring, Carl. "Dirty Work: How Hygiene and Xenophobia Marginalized the American Waste Trades, 1870–1930." *Environmental History* 9 (January 2004): 80–101.

———. "Recycling for Profit: The Evolution of the American Scrap Industry." Ph.D. diss., Carnegie Mellon University, 2002.

PERIODICALS

American City
American Metal Market
Associated Press
Atlantic Monthly
Century
Chemical Week
Cosmopolitan Magazine
Delaware County American
Electric Railway Journal
Engineering News
Fortune
Industrial Management
Institute Bulletin of the Institute of Scrap Iron and Steel, Inc.
Iron Age
Iron and Steel Engineer
Iron Trade Review
Manufacturer and Builder
Metal Industry
Minneapolis Star Tribune
Municipal Affairs
Nation's Business
New York Times
New York World Telegram
Pennsylvania Gazette
Pittsburgh Post-Gazette
Practical Magazine
Scientific American
Scrap Age
Smithsonian Magazine
Spokane Spokesman-Review
System
Technical World Magazine
Wall Street Journal
Waste Trade Journal

Index

About the Author

CARL ZIMRING received his doctorate from Carnegie Mellon University in 2002. He was an Environmental Protection Agency fellow from 2000 to 2002 and has been a visiting assistant professor at Michigan Technological University and Oberlin College and a lecturer in American history at the University of Canterbury, New Zealand.